DISCOVERING ENERGY

Opposite: An engineer makes final adjustments to a
vertical axis windmill. These machines, used mainly
for electricity generation, have the advantage of
needing no mechanism to turn them to face the
prevailing wind.

DISCOVERING
ENERGY
by Frank Frazer

Longman

Longman Group Limited
Longman House
Burnt Mill, Harlow, Essex, UK

First published in Great Britain 1982

Created, designed and produced by
Trewin Copplestone Books Ltd, London.

Set in Monophoto Rockwell Light by
SX Composing Limited, Rayleigh, Essex.

Separation by Scan Studios Ltd, Dublin.

Made and printed in Spain by Novograph S.A.

British Library Cataloguing in Publication Data
Frazer, Frank
    Discovering energy.
    1. Force and energy
    I. Title
    531'.6   QC73

    ISBN 0-582-25058-7

Contents

The World of Energy

Energy comes to us in many forms: as electricity to power our factories and warm our homes, as gas to cook our food, as petrol to drive our vehicles, and most vital of all as the heat and light we get from the Sun. In fact, the Sun is the Earth's ultimate source of energy. It not only radiates a plentiful supply of warmth, but has also helped to create coal, oil and the Earth's radioactivity, which are also vital sources of energy. Even the food we eat contains a dose of energy that originated in the Sun.

So what do we mean when we call all these things energy? A clue can be found in the origin of the word in the Greek language: it means action. And that is a very good description of energy in all its forms. Whenever energy is used, some sort of action takes place, so we can only begin to understand the nature of energy if we also look at what we mean by action.

We often think of action as some kind of bodily exertion. However, we also build powerful machines that are many times stronger than the human body. These can only function if they have a supply of fuel, just as we can only carry on working if we have a regular intake of food. Human beings and the machines they operate are therefore continually using up Earth's energy reserves and it is important to realize just how limited these are. Our present rate of consumption cannot carry on indefinitely, and energy is far too valuable to waste.

Fortunately there is no danger of the Earth suddenly running out of energy, because of the enormous doses it receives each day from the Sun. These set up complex natural processes by which many of our possible sources of food and fuel are renewed as others are exhausted, so we should have enough energy to last us for millions of years to come if we manage our resources wisely.

How we can best use and develop the Earth's vast energy reserves can only be understood with some knowledge of many branches of science. The inner nature of energy is studied in the science of physics, while chemistry reveals the way in which energy is released when different substances react with each other. The science of geology can explain how some of our most useful sources of energy come to be buried deep down in the Earth's crust and where valuable deposits of coal and oil are most likely to be discovered, while the basic principles of economics can tell us why the price we pay for energy is continually rising. The world of energy has many different aspects, and finding out about them is a fascinating journey of discovery which is of vital importance to us all.

Energy in Action

How does a jet aircraft lift its tremendous weight off the ground? What gives a runner the stamina to reach the finish line in a race, or a windsurfer the power to glide effortlessly through the water? How can a beam of light become so intensely hot that it will cut through metal? To answer each of these questions we need to talk about the transformation of one sort of energy into another.

The airliner gets its power from jet turbines. These create a high-pressure stream of very hot gases that push the aircraft forward as they leave the engine. In this way heat energy is transformed into movement, which is sometimes described as mechanical energy. However, this energy transformation could not begin without chemical energy, which the aircraft gets from fuel stored in tanks within its wings or fuselage.

Energy from the wind drives a windsurfer through the water.

The bodies of these track athletes are highly efficient machines that produce motion from chemical energy contained in food.

Heat energy is shown here in two guises; left, as the source of thrust in jet engines and, above, in a high-powered industrial laser.

Fuels can take the form of gases, solids or liquids. They release their stored energy as heat when they combine with oxygen from the air in the process we recognize as burning. Human beings rely on food for fuel, which contains energy-giving substances that our bodies can store until we need power to flex our muscles. We may not always be aware that heat is being given off when we exert ourselves, and our bodies do not burst into flames, but the beads of perspiration on our skin are a clue to what is happening. Perspiration is caused by the body's natural cooling mechanism which removes spare heat released as chemical energy is transformed into motion within the muscles.

The movement of the windsurfer has a somewhat different explanation. He is propelled along by a sail, which collects mechanical energy from the winds that sweep across the surface of the water. This energy has been produced as a result of heat from the Sun, which warms the Earth's surface and sets the air lying above it in motion. The Sun's heat comes to Earth as a form of radiant energy, which travels through empty space. When it reaches the Earth much of it is absorbed by the atmosphere and the surface of the land and sea, causing them to rise in temperature. In fact most of the heat we feel comes not directly from the Sun but results from the atmosphere being warmed by the Earth's surface. This is why some of the upper layers of the atmosphere are so much colder than the land beneath them.

Scientists used to think of the Sun as a mighty ball of fire burning up some sort of fuel. The centre of the Sun is certainly very hot – up to 15 million degrees Centigrade – but it is now known that this is because another kind of energy reaction is taking place in which new substances are continually being created as others are being destroyed. The reaction is known as a nuclear reaction and we are now attempting to imitate it on Earth to improve our energy supply. Scientists have calculated that the Sun has enough fuel to go on producing energy at its present rate for about five billion years.

On Earth man-made nuclear reactions are used to produce the form of power we know as electricity, which can be transformed into other kinds of energy such as heat, light and radio waves. Nuclear energy also has terrifying destructive potential when harnessed in atomic bombs. Electrical energy can also be used to produce laser beams, a type of radiant energy that can be concentrated at a narrow point, where the impact of so much power creates heat able to cut through metals like a knife slicing butter.

Measuring Energy

Before scientists could make a sophisticated study of energy, they had to find ways of measuring the work it does. From earliest times it was realized that horses, and some other domesticated animals like oxen, could do much more strenuous work than human beings. The average horse, for example, can haul ten times more weight than a human being, and even when engines began to take over work previously done by horses, the working ability of the horse was still used as a measure of energy.

In the eighteenth century, in the early years of the Industrial Revolution, James Watt, the Scottish pioneer of steam power, calculated the rate at which horses could move weight and used it to measure the working performance of steam engines. His unit of measurement, called horsepower, showed that some of the early steam engines could do little more work than a single horse, but nowadays engineers can build machines that will yield an output of power

Horses, such as these pulling a plough on a farm, were once one of our chief sources of energy. Their strength became a unit of measure, the horsepower, that has survived into the industrial era.

equal to the combined working ability of more than one million horses. Even a compact car, using an engine fuelled by petrol, will probably be as powerful as 100 horses, or 1000 human slaves. Scientists have even calculated that modern standards of living in industrialized countries demand an input of energy equivalent to the services of a hundred slaves for every member of the population.

One of the most basic forms of energy is heat. The amount of heat in a substance depends upon the movement of the millions of minute particles of which matter is composed. Temperature depends upon the intensity of this movement, and can be measured very simply by using a thermometer. In its commonest form a thermometer consists of a glass tube containing a liquid that expands when heated and contracts when cooled. A scale marked on the tube relates the expansion to standard units of measurement, sometimes degrees Centigrade or Celsius, at other times degrees Fahrenheit.

Temperature measurements have helped to show the relationship between heat and other forms of energy. In the nineteenth century James Prescott Joule, the son of an English brewery owner, succeeded in measuring how much heat could be obtained from mechanical energy. Joule gave his name to one of the units that scientists throughout the world use to measure all forms of energy. Scientists define the joule as the amount of energy needed to move a mass of one kilogramme through one metre with an acceleration of one metre per second per second.

Heat can also be measured in a unit called a calorie. This is distinguished by its small c from the familiar dietician's Calorie, which is equal to 1000 calories. A calorie is defined as the amount of heat required to raise the temperature of one gramme of water through one degree on the Centigrade scale, and experiments show that it is equivalent to 4.184 joules.

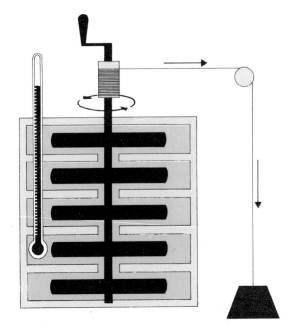

James Joule's apparatus for correlating heat energy and mechanical energy consisted of paddles rotating in a container of water. He turned the paddles by a weight on the end of a string suspended over a pulley. A thermometer showed a similar rise in water temperature every time the same weight fell by the the same distance.

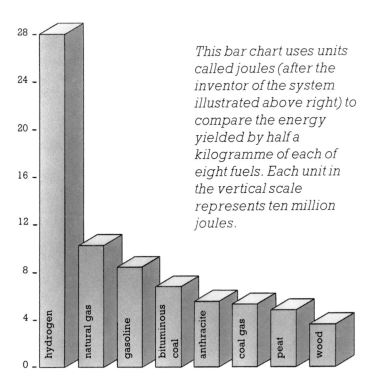

This bar chart uses units called joules (after the inventor of the system illustrated above right) to compare the energy yielded by half a kilogramme of each of eight fuels. Each unit in the vertical scale represents ten million joules.

Both joules and calories are very small measures of energy and when scientists deal with large units, they add prefixes such as 'kilo' meaning a thousand and 'mega' meaning a million. So a kilojoule is 1000 joules and a megajoule is 1 000 000 joules.

Comparisons are frequently made of the amount of heat given off by different fuels. For example, an average tonne of oil has been found to yield about the same amount of heat as one and a half tonnes of coal or the burning of more than 11 600 cubic metres of natural gas. When comparing the energy value of different fuels, it is common to convert the actual volumes into the equivalent volume of either coal or oil. For example, the United States in 1979 needed supplies of energy that were equivalent to burning nearly 1.9 billion tonnes of oil, but less than 900 million of that total was actually met by oil. The rest was provided by getting the equivalent energy from other fuels such as coal and natural gas.

Energy in the Universe

The natural force that keeps our feet firmly on the ground as the Earth spins around on its axis is known as gravity, and describes the attraction that all objects in the universe exert on one another, with a strength that depends on their masses and distance apart. Scientists still do not know the true nature of gravity but studies have shown that it exists throughout the universe and is a fundamental force of nature.

The Earth's natural gravity can be overcome by using another form of energy to counteract its attraction. This is what happens if we jump up in the air: as the energy in our movement is used up, we are pulled swiftly back towards the Earth. Astronauts on the Moon can take much bigger leaps with the same amount of muscle power because the Moon has less mass than the Earth and exerts a less powerful gravitational pull. However, even there anyone trying to break free from gravity would need far more energy than can be provided by muscle power.

Enough chemical energy is concentrated in some fuels to enable space rockets to overcome the gravity which normally makes objects fall back to the Earth's surface. The release of energy in the fuel propels the rocket with such force that its speed at a certain height above the Earth balances the force being exerted by gravity. At that point the spacecraft will begin circling the Earth without the help of any energy other than the natural force of gravity, just as the Moon circles the Earth. The speed at which a spacecraft must travel to achieve this condition, known as orbital velocity, is typically about 40 000

The structure of atoms can be pictured as a minute solar system with electrons "orbiting" in three dimensions around a central nucleus composed of protons (red) and neutrons (blue) or, in the case of hydrogen, a single proton.

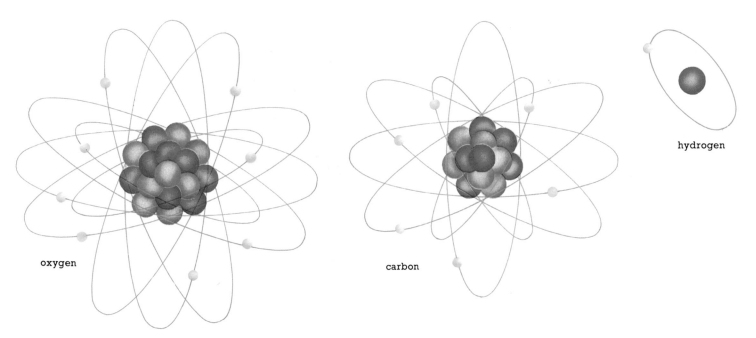

oxygen

carbon

hydrogen

kilometres per hour and can be calculated by using a formula based on studies of gravity.

Although the Earth has its own gravitational pull, it is also affected by the pull exerted by the Sun and the planets. And the Moon, which is held in orbit by the Earth's gravity, exerts enough pull to cause water in lakes and oceans to move away from the Earth's influence, creating the rhythmic rise and fall of tides. The Earth itself is one of a group of planets that move in orbit around the Sun, which is the most powerful gravitational force in our part of the universe. The Sun exerts such a strong pull that it even holds in orbit the tiny planet of Pluto more than 5.6 billion kilometres away.

Forces similar to gravity provide a kind of 'glue' which helps to hold together every particle of matter in the universe. There are still many mysteries to be solved about the nature of matter but we do know that it is made up of tiny packages of energy called atoms. A single atom is so small that it can only be seen through the most powerful electron microscopes. Indeed, the full stop at the end of the last sentence has been formed by ink containing more atoms than there are people in the world. It would take a collection of probably one million million million million atoms to weigh one gramme.

In spite of the solid appearance of the world about us the atoms of which it is composed each consist of two or more particles separated by space. The energy which binds these atomic particles together can be harnessed to provide us with heat, light, and many other forms of power.

The 3.5 million kilogrammes of thrust needed to lift a 3000-tonne space rocket out of the atmosphere at Cape Canaveral indicates the strength of the Earth's gravitational pull.

Atomic Bonds

Powerful forces keep electrons linked to the protons and neutrons that form the nucleus of the atom and also enable atoms to combine with one another. Some substances are made up of identical atoms, each containing exactly the same number of protons and electrons. Such materials are called elements because they are the most elementary, or simple,

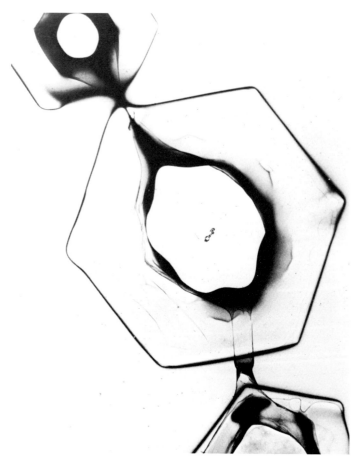

These rigid symmetrical plates of ice crystals, less than one millimetre across, are a result of the firm bonds that form between water molecules when subjected to temperatures below 0° Centigrade. Their hexagonal form is determined by the shape of an electrostatic field that surrounds each molecule.

form of matter. However, there are several million other substances known as compounds which are made up of groups of two or more types of atom. These groups of atoms, called molecules, are held together by forces arising from the electric charges of electrons. Bonding occurs when atoms are able to exchange or share electrons from their outer orbits.

Sometimes atoms of different kinds join with each other and give out some of their energy in the form of heat. This happens when hydrogen, the simplest element with just one proton and a single electron, combines with oxygen, a more complicated element containing eight electrons. Two atoms of hydrogen become attached to a single atom of oxygen by sharing an outer ring, or shell, of electrons. This creates the familiar liquid we call water.

Some of the energy which holds atoms together in their separate states is no longer needed when they combine, and is released as heat. This process often takes place when materials combine with oxygen, and is known as combustion. We recognize it more readily as burning, which takes place when a suitable fuel is used to provide heat energy.

Burning fuel always forms new substances made up of a combination of oxygen and other atoms. The bonds between atoms in these new molecules are very strong and can only be broken down by replacing the energy that has been taken out. In the case of water this would mean applying heat of more than 2700 degrees Centigrade, so strong are the bonds between the atoms of oxygen and hydrogen. On the other hand, the bonds that link individual molecules are relatively weak and comparatively little heat is needed to break them. For example, if heat is applied to a liquid its molecules begin to move rapidly about and collide with each other. Forces of repulsion then cause them to spring apart again and if the collisions are violent enough the bonds of attraction may be broken. This can be seen when we boil a kettleful of water. The molecules break free of the liquid and float upwards through the spout of the kettle as a cloud of water vapour. The vapour behaves very differently from the water, but it is still

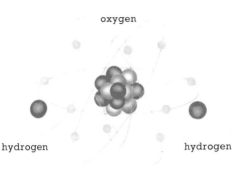

This schematic drawing of a water molecule shows that it contains two atoms of hydrogen linked to an atom of oxygen by sharing an outer ring of electrons.

Burning is an irreversible chemical process. Even if we replace the energy lost as these lumps of coal are transformed into ash, we will never be able to reconstitute the lumps of coal.

composed of atoms of hydrogen and oxygen firmly bonded together. Only the bonds between individual molecules are broken by the heat.

If we reverse the process and start cooling vapour the molecules soon turn back into water in its familiar liquid form, and if the cooling continues the water eventually freezes into a solid block of ice. When this happens the bonds between the molecules become so strong that they form a rigid substance.

Since heating weakens the bonds that keep molecules held closely together, the hotter a substance is the more it will expand. Expansion in heated metals and contraction in cooling metals causes problems for engineers who have to leave gaps between sections of bridges with metal girders so there is space between individual sections to prevent buckling or cracking caused by the force of expansion in hot weather and contraction in cold.

Engines

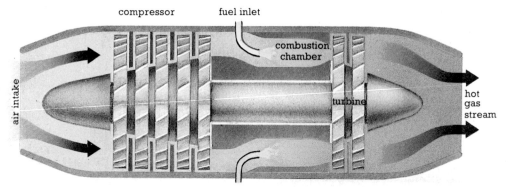

compressor   fuel inlet

air intake

combustion chamber

turbine

hot gas stream

Air is sucked into the front of a turbojet engine and compressed by alternate sets of rotating and stationary blades before being mixed with fuel. The air–fuel mix is ignited in the combustion chamber to provide thrust.

Mechanical energy can be produced by making use of the expansion which takes place when substances absorb heat. For example, at normal atmospheric pressure steam occupies about 1100 times more space than the same number of molecules of water would need if they remained in liquid form. If we tried to put a cork on the spout of a boiling kettle, we would soon find it being shot into the air by the pressure of the expanding molecules trying to escape and find more space.

From the earliest times attempts have been made to capture the potential energy of steam. Around 1629 Giovanni Branca, an Italian architect, experimented with a machine that directed a jet of steam from a container of boiling water towards a cogwheel, producing rotary motion. Various attempts were made to construct steam engines during the late seventeenth and early eighteenth centuries and in 1769 James Watt perfected a design that was successfully used to pump water out of coal mines.

In a steam engine based on Watt's design mechanical power is produced by feeding the steam into a cylinder closed by a sliding piston. The pressure of the steam forces the piston from one end of the cylinder to the other. But if the steam is then cooled by passing it through a valve into a separate chamber, it will rapidly contract and create a partial vacuum in the cylinder that will draw the piston back. The push–pull motion of the piston can be converted into power that will drive machinery, and this method produced the main source of power for industry during the

nineteenth century, which became known as the Great Age of Steam.

Another highly efficient way of converting steam to mechanical energy was discovered toward the end of the nineteenth century. A jet of steam is directed at a set of fan-like blades causing them to turn. If the steam is directed through a chamber containing many sets of blades linked to a central shaft, it will produce substantial amounts of power, and turbines based on this principle are now the chief method of turning steam into mechanical energy.

Cylinders and pistons are still used in the internal combustion engine, so called because the actual burning of fuel to produce energy takes place within the engine and not in a separate boiler as in steam-driven engines. Internal combustion engines, usually fuelled by petrol, are installed in most road vehicles in use today. A mixture of petrol or diesel oil and air sucked into the cylinders is ignited by an electrical spark or the heat of compression in the cylinders to produce an explosive force as the air and fuel unite. This force pushes the piston forwards to produce mechanical energy. Car engines have several cylinders and pistons linked to a central drive shaft which provides a rotating motion to turn the wheels.

In both steam and petrol engines, some energy is lost in the form of heat generated by friction between the moving parts. This can be avoided by producing motion directly from high-pressure gas. One way of doing this can be demonstrated by filling

A 604-tonne Big Boy locomotive, one of the largest steam engines ever built. It used up to 50 tonnes of water and 22 tonnes of coal an hour.

a balloon with air until the pressure is much greater than the air around it. If the inflated balloon is released, it will fly off as a result of the compressed air escaping through the narrow outlet at the neck. The force of the air rushing out will continue to push the balloon along until all the air left inside is at the same pressure as the atmosphere. This same principle of thrust is used in jet engines and space rockets.

Jet engines produce thrust by burning fuel along with oxygen to produce a high-pressure flow of hot gases. Space rockets travelling beyond the Earth's atmosphere have to carry their own oxygen supply but aircraft can draw oxygen from the air they are flying through. Most modern aircraft have a type of engine known as a turbojet which can suck in and compress the more limited amounts of oxygen at high altitudes, ensuring that the fuel will still burn efficiently and give the maximum amount of thrust.

Four stages in the operation of a car petrol engine:
induction, *as a petrol–air mix is sucked into the cylinder;* compression *as the piston rises within the cylinder;* ignition *of the fuel by an electric spark causing an expansion of gases that pushes the piston down, and* exhaust *as the piston rises again and forces out waste gases.*

induction

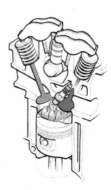

compression

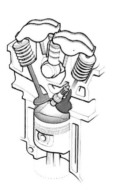

ignition

exhaust

Earth's Energy System

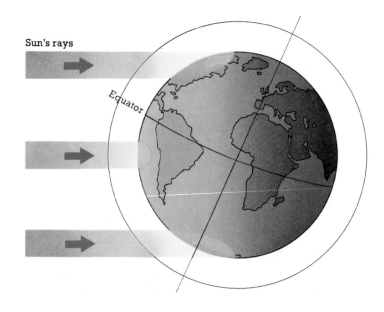

Sun's rays

Equator

The world in which we live has its own highly efficient energy system. It is fuelled by the Sun which each day pours down upon the Earth more than 10 000 times as much energy as we need. This energy, known as solar radiation, provides us with heat and light, and it helps to create the winds, which are another useful source of power. Solar radiation also causes the clouds and rain which reduce the amount of sunlight that reaches the Earth's surface, but even the rain can provide us with a source of energy.

It took scientists many centuries to puzzle out how the Earth's energy system worked. The secret lies in the fact that the Earth is constantly moving in relation to the Sun. This means that the amount of solar radiation falling on any part of the Earth's atmosphere varies from hour to hour and month to month, and that different places receive varying amounts of sunshine at any one time. The resulting pattern of hot and cold areas within the Earth's atmosphere causes the movement of air and water we recognize as wind and rain.

The parts of the Earth which receive most solar

Because of the Earth's curvature, areas near the Equator receive more than twice as much solar radiation as those near the poles. Sunlight arriving in polar regions also has to pass through more of the Earth's atmosphere, which absorbs its heat and further reduces the amount of energy reaching the surface. The Earth is shown here in December, with the northern hemisphere tilted away from the Sun and the southern hemisphere tilted towards it.

radiation are near the Equator, where the Sun appears high in the sky at midday throughout the year. This is where the most direct rays of sunlight strike the Earth's surface, unlike areas toward the polar regions where radiation falls at a more oblique angle and so is spread over a greater area. Temperature differences also occur due to the tilted axis on which the Earth spins, causing the angle

This diagram shows the working of our planet's 'weather machine', which uses more than 50 per cent of the solar energy reaching the Earth's surface. One of the key mechanisms, evaporation, takes up about 0.025 centimetres of water over the globe daily.

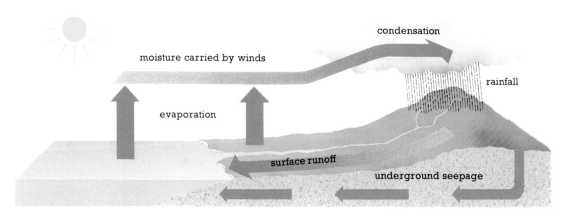

condensation

moisture carried by winds

rainfall

evaporation

surface runoff

underground seepage

Volcanic lava reaches the surface in a dramatic release of energy from the Earth's interior.

at which areas receive solar radiation to change as the Earth makes its annual journey around the Sun. This is the reason for the seasonal variations in temperature and hours of sunlight, which become more marked toward the poles.

When the Earth's atmosphere is heated by the Sun it behaves rather like air being warmed by a radiator in a big room. Air heated by coming in contact with the radiator expands and rises to push away colder air occupying the space above. This cold air is forced to travel along the ceiling, down the opposite wall and along the floor until it too is warmed by the radiator. Then it will start to rise, pushing away air that has now cooled through losing heat picked up when passing the radiator earlier. And so the cycle will continue, forming a convection current, as long as there are hot and cold areas in the room.

However, the pattern of convection currents in the Earth's atmosphere is much more complex. Most of the warm air results from heat generated by concentrated solar radiation falling in the regions nearest to the Equator. As the warm air rises, colder air moves in from less hot areas to take its place but the air currents are deflected by the Earth's spin,

producing a complicated pattern of wind systems.

The effect of solar radiation also creates rainfall. Sunshine falling on oceans, seas, and lakes causes surface water to evaporate and rise to form clouds which are carried off by the winds. The water vapour changes back to liquid and falls as rain when the air that carries it is cooled, for example by rising to cross a mountain range. This is why mountainous regions are among the wettest places on Earth.

Not all of the Earth's energy originates directly from the Sun. The movement of tides, which also contain enormous amounts of power, is caused by the gravitational pull of the Moon. Energy even reaches us from far below the surface of the Earth, which has an intensely hot core (3800 degrees Centigrade). We are not usually aware of this heat because we receive up to 3600 times as much energy from the Sun shining above. Yet over millions of years both sources of energy have helped to build up hidden stores of fuel upon which we now rely heavily.

The Rock that Burns

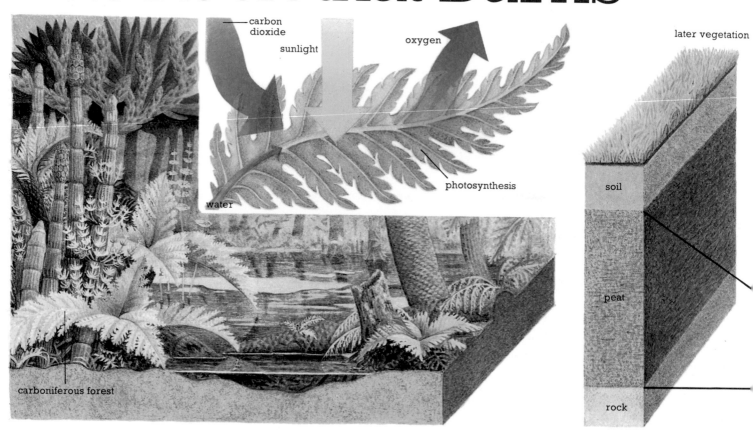

carbon dioxide

sunlight

oxygen

later vegetation

photosynthesis

water

soil

carboniferous forest

peat

rock

Much of the coal we burn today had its origin on the surface of our planet more than 250 million years ago in the Carboniferous Period. We can only imagine what conditions were like that far back but we can be certain of one thing: there were large areas of luxuriant vegetation to provide the raw material for the formation of coal.

When the Earth had finally cooled down after its early fiery period, the first plants began to grow, and much of the land became covered with forest. The leaves drew in carbon dioxide from the atmosphere while the roots took up water from the soil. With the aid of energy from sunlight, the carbon dioxide was fused with the hydrogen from the water to produce carbohydrates, which are a form of stored energy. Finally, surplus oxygen was given off into the atmosphere. This process, known as photosynthesis, is used by all green plants to make the food on which much human and animal life depends.

In time, as plants died, carbohydrates combined with oxygen again and gave off unused energy in the form of heat. However, this process was sometimes prevented by heavy rainfall and low evaporation which waterlogged the soil and sealed the dead vegetation away from the oxygen necessary for normal decay to take place. As more and more layers of material accumulated this resulted in the formation of a substance called peat, which resembles a closely packed layer of rotten wood. Peat deposits are usually between one and three metres thick and can be found today in many parts of the world. In Scotland, Ireland, Scandinavia and Russia they are often dug out in brick-sized blocks to provide a slow-burning, smoky fuel.

Frequently peat has been buried under many layers of sediment in the processes which formed the

Over millions of years vegetable matter, formed by photosynthesis, has turned first into peat and then become compressed by soil and sediment to form deposits of rough and impure coal called lignite. The

lignite itself has been further compressed to produce purer grades of coal called bitumen and anthracite. Later Earth movements have sometimes contorted coal seams, causing great difficulties to miners.

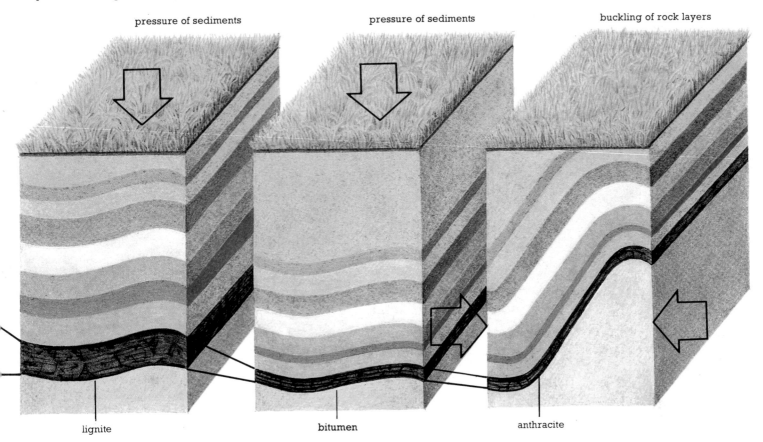

pressure of sediments   pressure of sediments   buckling of rock layers

lignite   bitumen   anthracite

Earth as we know it today. Each new deposit of sediment made the peat sink farther below the surface until eventually this trapped store of energy ended up hundreds or even thousands of metres underground. The greater the depth, the more layers of sediment pressed down upon it from above.

The pressure and the heat in the depths of the Earth's crust have gradually transformed the peat into a rock which we recognize as coal. A layer of peat that was 18 metres thick under its first covering of sediment might now be compressed into a coal seam only one eighth that size. During the process its chemical structure will have changed into a more concentrated compound of hydrogen and carbon known as a hydrocarbon. And because coal is derived from living things that flourished on the Earth's surface millions of years ago it is called a fossil fuel.

Generally, the coal seams which have been most tightly compressed provide fossil fuel giving off the greatest intensity of heat. Coal seams that have been under least pressure produce a fuel known as lignite or brown coal. Brown coal forms less than a quarter of the world's total known coal deposits. The rest consists of anthracite, or hard coal, and bitumen, or soft coal. Of these, the former contains the most carbon and burns with the cleanest flame while the latter generates more smoke because of its greater moisture content.

We usually have to dig deepest into the Earth's crust to extract the most valuable coal – but not always. Coal seams have sometimes been pushed sideways and even upwards by movements deep within the Earth's crust. This results in deep-lying coal deposits being forced back towards the surface to form an outcrop.

Deep Mining Today

Coal was being burned as a fuel in Europe nearly 4000 years ago but it was not until the nineteenth century, when the Industrial Revolution brought fresh demands for coal and new techniques for its extraction, that coal mining developed on a large scale. Mining reached a peak in the early 1920s, when at one time it employed over 700 000 workers in the United States alone, but then began to decline as industry made increasing use of cheap supplies of newly discovered oil and natural gas. Now, with the continued availability of these fuels in serious doubt, industry is showing renewed interest in coal as a source of energy. This has stimulated fresh developments in mining technology, and a deep mine of today is very different from the cramped and dangerous mines of a hundred years ago.

The development of a coalfield today is a carefully planned operation. It begins on the surface, above the area where geological surveys have shown that there could be coal deposits. The mining authority first carries out a series of test borings to find out the extent and thickness of any coal seams. The results help to provide a three-dimensional map of the coal seam and enable mining engineers to decide the best way to develop the project. Access shafts are then sunk to the depth of the seam and large, roomy tunnels for the railway system are driven on either side of the coal faces that will later be worked.

Careful planning and mechanization enables many coal seams to be worked by a system known as longwall mining. This is more efficient than the traditional room and pillar technique, which sometimes leaves more than half of the coal behind to support the upper layers of rock. The longwall method enables coal faces up to 180 metres in length to be worked by rotary shearers which go backwards and forwards along the coal face between the two access tunnels. The coal is removed by conveyor belts and the roof above the machinery is held in place by steel supports

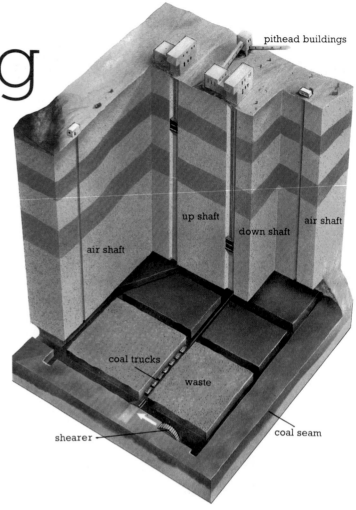

This cutaway diagram shows the layout of a modern longwall mine. A shearer moving backwards and forwards along the coal face feeds a line of trucks, which transport coal to access shafts, while the space created by the coal's removal is allowed to fill with rubble in a controlled collapse of the roof. Ventilation shafts ensure a constant passage of fresh air through the working area.

which are moved forwards as the cutting progresses. The carefully planned operation allows rock above the seam to fall gradually into the space left after the coal is removed, while the roof above the access tunnels at either side of the face is reinforced and supported to make sure there is a safe passage in and out of the mine.

Safety is of paramount importance in modern mining. The layout is planned to ensure a continuous one-way flow of air from the surface to remove

dangerous gases such as methane and carbon monoxide and provide a fresh atmosphere for the miners. All electrical equipment used underground has to be built to standards that ensure that no stray sparks will ignite any of the methane and carbon monoxide gases that escape from coal seams when they are cut. Coal dust, too, is an explosive hazard. It is also dangerous because, if inhaled by miners, it can lead to a disease called pneumoconiosis or black lung disease, the chief symptom of which is acute shortness of breath. Most countries now enforce strict regulations to limit the amount of dust allowed in mines, and modern coal-cutting machines have built-in water sprays which help to reduce dust amounts.

The only way to eliminate mining hazards completely is by greater use of automation. Already underground transportation of coal can be controlled

Advancing along the coal face at up to six metres per minute, this shearer uses tungsten carbide blades to tear out 9000 tonnes of coal in one day. Water jets and a rotating speed of only 30 revolutions per minute help to prevent choking clouds of coal dust.

from the surface by an operator watching closed-circuit television screens. For example, robot miners controlled by computer may one day take over the tasks of operating and repairing machinery at the coal face. Although these developments require the investment of vast sums of money, the use of advanced mining technology will be worthwhile if it makes possible the large-scale exploitation of the only natural resource that is a proven alternative to our shrinking supplies of oil and natural gas.

Surface Mining

Where coal seams lie close to the surface, it is often possible to remove the soil that covers them and extract the coal by quarrying. This technique is known as open cast or strip mining, because it involves stripping away the 'overburden' of soil before the operation can begin. It is particularly suited to exploiting the large areas of coal deposits beneath the North American prairies.

Colorado, Wyoming and Montana are among the states in the western United States which contain huge coal reserves suitable for strip mining. In these areas seams have been thrust upwards by movements in the Earth's crust and now lie buried under just a thin covering of overburden which can easily be shifted by earth-moving machinery and dumper trucks similar to those used in road building and building-site excavation. After the removal of the overburden, coal is extracted by massive cutting and digging machines which gnaw at the seams. Far fewer men are needed for this type of mining than for underground operations and some strip mines in the United States can achieve production rates for each person employed that are more than three times better than the average for underground workings.

Because operations take place above ground, strip mining is comparatively safe, with one half the injuries in the United States and about ten per cent of the number of deaths than would be caused by recovering the same amounts of coal from underground. But strip mining has its disadvantages. The removal of the overburden can permanently disfigure the landscape. Although deep mining produces piles of waste that are a notorious source of pollution and are visually disturbing, its impact is not nearly as obtrusive as the ugly cavity left after strip mining. In addition, there is often a danger of local rivers becoming polluted during mining operations. This is because the removal of the clay in the overburden allows ground water to seep through coal seams, which usually contain a percentage of sulphur. The resulting sulphuric acid eventually drains into the rivers, upsetting their ecological balance.

Environmental drawbacks of this sort often create local opposition to strip mining operations, even in regions where there are rich coal deposits near the

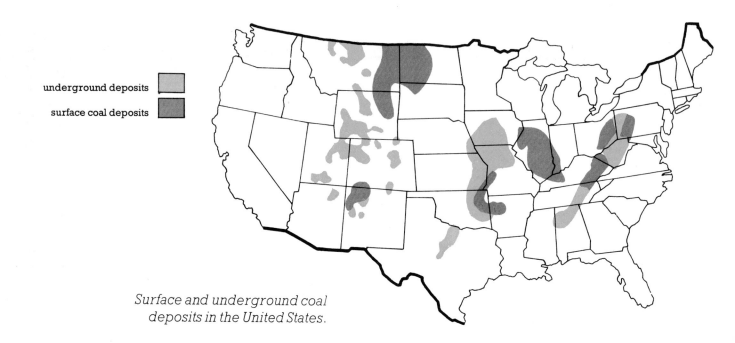

underground deposits

surface coal deposits

Surface and underground coal deposits in the United States.

Spoil heaps and a complex system of access roads leave an ugly blot on the landscape at this Welsh strip mine.

An example of the restoration of land disfigured by mine workings in Scotland. The picture below left shows the hill of waste material that used to dominate a village that fell into decay when the mine closed. Below right is a picture of the same area today, now a gently contoured landscape with only the road and the lamps left to recall the past.

surface. When strip mining is permitted mining companies are made to restore the landscape where the coal has been removed. At large mining sites coal extraction and land restoration are carefully co-ordinated: earth from the first area to be worked will be carefully piled nearby so that it can be used to fill the cavity when the coal has been removed and the diggers have moved on to another area where the process will be repeated. Attempts are made to restore the land to its original contours, and grass and trees may be planted to make the restored area blend in with its surroundings. In some places former strip mining sites can even be used for agricultural purposes.

Coal to the Consumer

Coal is often many kilometres from the industrial areas that need it for fuel. Since it is particularly inconvenient to handle, the cost of transporting coal is high, and may amount to a high proportion of its eventual price.

One method of cutting transport costs is to use specially designed trains. In the United States, for example, unit trains composed entirely of coal cars normally carry between 10 000 and 15 000 tonnes of coal directly from the mine to the consumer. The world's longest-ever train was unit train that carried coal 255 kilometres from Iaeger, West Virginia to Portsmouth, Ohio. Over six kilometres long, its 500 cars contained a total load of 32 000 tonnes. In Britain an automated system is used which enables more than 1000 tons of coal to be loaded or discharged in less than half an hour – so rapidly that trains can be kept moving throughout the operation.

The cost of transporting coal overland can also be reduced by using pipelines. The coal is crushed – or pulverized – at the mine to produce a fine dust. This dust is mixed with water to form a thick black slurry which is pumped through to the place where the coal is needed – perhaps hundreds of kilometres away. There the coal particles are separated from the water by a combination of natural settling, filtration, and spinning in a centrifuge, and then dried to produce a fuel suitable for burning in furnaces.

However, pipelines are not suitable for transporting coal between continents, and in the future increasing amounts of coal may be carried by sea in specially constructed cargo ships. Although shipping coal is expensive, larger ships can cut transportation costs, and modern ports help to keep prices low by allowing cargoes to be loaded and unloaded more rapidly. Australia has vast coal deposits and, with new methods of carrying cargoes, could become a leading exporter in years to come.

Another way of minimizing the cost of transport is to convert the coal into a form of energy that can more easily be sent over long distances. Some mines feed their coal onto conveyor belts that carry it straight into power stations sited near the top of the shaft. This system cuts out the need to deliver coal by road, rail or barge and it is more convenient to send energy from the mine in the form of electricity.

Coal nearing the surface at Longannet Colliery, Scotland, where the world's largest conveyor belt extends through nine kilometres of underground workings. Conveyor systems are one of the simplest ways of transporting coal over short distances and, as here, provide a continuous link between mine and power station.

Coal can also be converted to gas and distributed by pipeline. One day it may even be possible to turn coal into gas without even bringing it to the surface. This would be done by carefully controlled underground burning. Oxygen would be piped down to maintain the burning process and would combine with carbon and hydrogen to form a gas that could be collected through another borehole driven into the seam. This could be a useful way of extracting energy from seams that are too thin or deep-lying to be mined by conventional means.

Various other conversion processes may one day be carried out in a new type of enterprise known as a coalplex, which would contain a cluster of processing plants grouped around a coal mine. Each plant would take coal from the mine and turn it into electricity, gas, or liquid fuel. The coalplex would distribute many of the fuels used by industry today – except the coal itself!

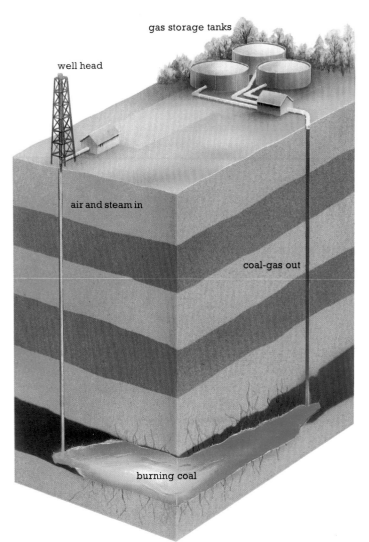

A map of world coal shows extensive deposits away from established mining regions in Europe and the United States. Large reserves in China, the Soviet Union and Australia are now gaining fresh importance as oil price rises make transport of coal to distant consuming regions economically worthwhile. Actual reserves may be much greater than indicated here because exploration is still in progress.

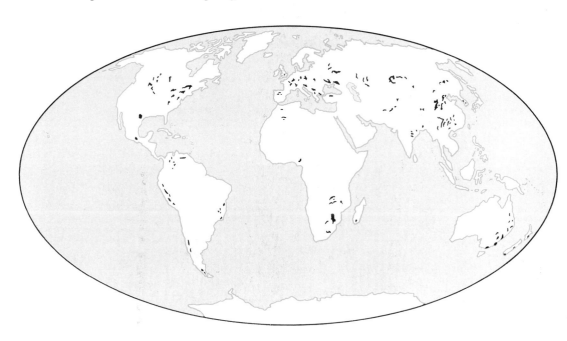

Coal beds that are too difficult and costly to mine can be made to give up their energy in the form of gas. Engineers ignite the bed with an electronically detonated charcoal fuse and feed the flames with a continuous supply of air and steam. Coal gas is then collected from a borehole. Pilot coal gasification projects are now under way in France, Belgium, Germany and the United States, and commercial operations could begin by 1990.

Coal in Use

For centuries coal was regarded mainly as a fuel to be set alight and burned in open fireplaces as a source of heat. Because it gave off a greater intensity of heat than wood, it was well worth the effort of digging deep into the Earth to recover coal and transport it to homes and factories.

But simply burning coal in an open fireplace is a very wasteful way of using this valuable source of energy. Much of the heat escapes up the chimney, which has to be installed to take away dangerous gases such as carbon monoxide that are produced as coal burns. These gases also cause pollution when they are discharged into the atmosphere, impairing the quality of air we breathe.

Nowadays coal is often washed after extraction to remove impurities that would cause pollution when it is burned. This does not entirely overcome the problem because some of the pollutants are chemically bonded to the hydrocarbons in coal and will only break loose when the fuel is set alight. However, if tall chimneys are installed above coal-burning factories, gases escaping into the atmosphere will more readily be dispersed by fast-moving upper air currents.

Sulphur is one of the worst pollutants contained in coal. During the combustion process, it combines with oxygen to form sulphur dioxide gas, which will react with rainwater to produce sulphuric acid. In

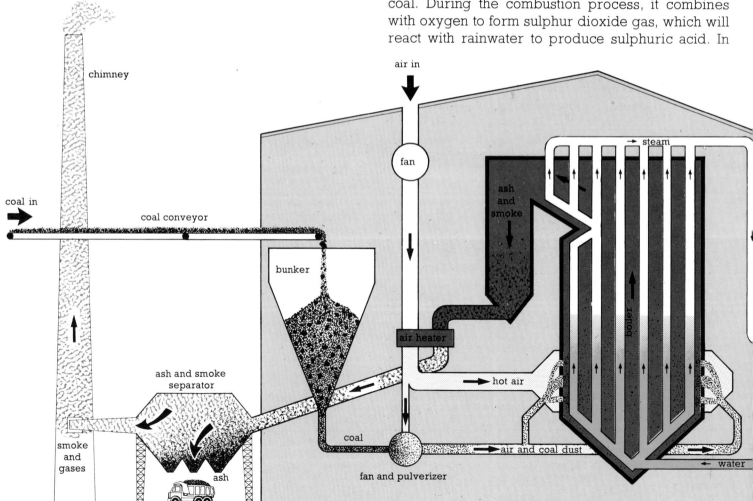

Coke, made from bituminous coal, emerging from the oven at a modern plant. The world's steel-making industry consumes 1.5 tonnes of coke for every tonne of steel produced.

In a modern coal-fired power station, a mixture of coal and air is burned in the furnace through which water is piped to pick up the heat. This produces high-pressure steam that drives a turbine connected to a generator.

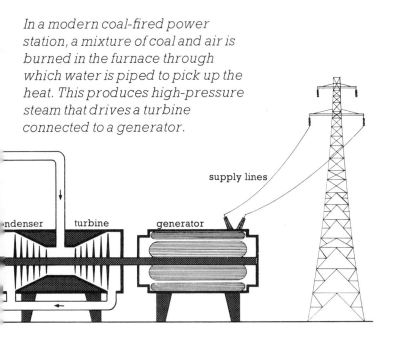

areas where sulphur dioxide is emitted from coal-burning installations this acid rain can kill fish and damage vegetation if it is allowed to build up in high concentrations in lakes and the soil. However, the gases given off by coal can now be treated by bubbling through chemical solutions to remove much of the sulphur.

Coal can be processed to produce a much wider variety of substances than the coke and gas with which it is usually associated. Carbonization, a process in which coal is baked in an airtight oven to yield coke, tar and gas, is the beginning of a whole series of refining techniques which can yield products such as pharmaceuticals, plastics and dyes. Some of today's most popular perfumes started life as far from sweet-smelling tar acids, and even the food we eat contains substances derived from coal tar in the form of preservatives and artificial vitamins. Another refining process, hydrogenation, involves treating coal under pressure in combination with oil and hydrogen. This produces gases which can be distilled into a range of products similar to those obtained by carbonization.

Although the amount of coal burned in open fireplaces has declined in most countries in recent years the techniques developed for modern coal-burning power stations have led to the design of domestic stoves that also give great efficiency in converting the fuel into heat. Even if coal fires are completely abandoned as a form of heating most homes will probably continue to benefit from energy supplies derived from coal in the form of either electricity or gas.

Coal is so versatile it can even be turned into a liquid fuel. This has been done on a commercial scale in South Africa for nearly 30 years, and although it requires costly processing, research work is currently under way in a number of other countries, including the United States, in an attempt to find less expensive ways of breaking down the solid hydrocarbon composition of coal. This would yield other combinations of the same two elements of hydrogen and carbon in liquid form. Some day it may become quite common for motorists to fill their fuel tanks with fuel that has been derived from some of the world's abundant reserves of coal.

Oil in Store

Although oil deposits have been formed by geological processes similar to those that created coal seams, the origin of oil itself is more difficult to discover. It probably derives from stores of carbon and hydrogen that built up in the plants and tiny animals which lived in the shallow seas covering many parts of the Earth 400 million or 500 million years ago. The remains of these creatures and plants eventually sank and mixed with mud on the sea bottom. As new layers of silt were washed into the seas by rivers, the mixture was sealed off from the oxygen necessary for natural decomposition to take place and release the stored energy back into the atmosphere in the form of heat. Instead, the pressure and heat caused by the build-up of new upper layers of material gradually turned the mud into a firm rock and, over millions of years, forced out the carbon and hydrogen. These components have again been transformed by heat and pressure to become the fossil fuel we call oil.

Oil, like coal, is a hydrocarbon fuel. The particular combination of hydrogen and carbon determines whether the fuel takes the solid form of coal or the liquid form of oil. Oil has a higher proportion of hydrogen than coal and this makes it lighter in weight. One tonne of oil has about the same energy content as one and a half tonnes of coal.

Another name given to naturally occurring oil is petroleum. The word is Latin for 'rock oil' and is an apt description. But the rock in which oil is found has a special characteristic: it is composed of millions of minute grains compressed against each other with space in between. The liquid oil has collected in this space. An oil-bearing rock is often called an oil reservoir, but this is a misleading term if it suggests that the fossil fuel is stored in huge underground caverns which will be left void when the oil is removed. Once oil has been extracted, reservoir rock will look much as it did before.

Unlike coal, in which the stored energy actually takes the form of rock, oil is an entirely separate substance from the rock in which it was formed. This, and its liquid form, means that it is very mobile and tends to move upwards as the heat in the Earth's crust causes it to expand. Rising oil frequently penetrates surrounding rock formations and continues its upward journey until it meets a layer of rock which is so tightly compressed that there is no passage through. Such a barrier might be produced by movements in the Earth's crust that have caused hard, impenetrable rock to come against the porous layers through which oil is moving. When this happens, the oil will become trapped in a formation that we recognize today as an oilfield.

Usually water has also been trapped underground in the same layers of rock and some of the hydrogen and carbon atoms from the oil have combined to form a mixture of gases. Because the gases are lighter than the oil, they gradually bubble through it and gather at the top of the reservoir. Similarly, oil is lighter than water and will float into the pore space above it. However, in most oilfields, there will not be enough space for all three substances as they expand in the high temperatures deep within the Earth's crust. This is why pressure builds up and drives oil to the surface when a well is drilled.

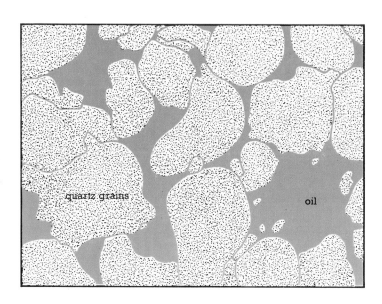

The cavities between grains of quartz in sedimentary rocks provide space in which oil deposits can accumulate.

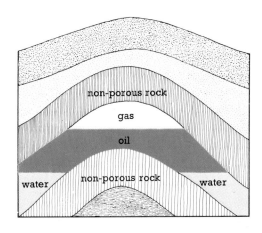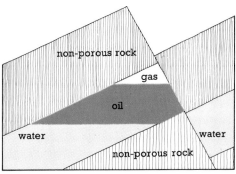

These diagrams show typical geological formations in which oil traps can occur. In both cases the oil lies between a layer of gas above and water below.

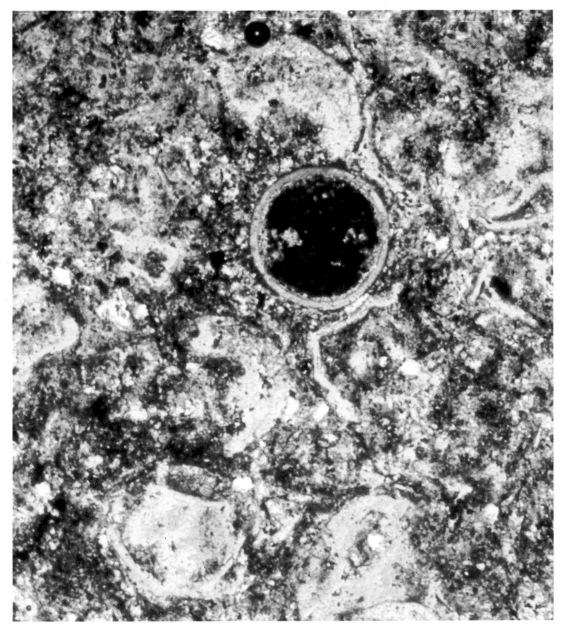

A slice of Tasmanite rock reveals globular algae 0.13 centimetres in diameter whose crushed bodies form a substance called kerogen, half way to becoming petroleum. The black sphere near the centre is the body of an organism that has been infilled with iron sulphide at an early stage in rock formation.

Finding Oil

The world's first recorded oilfield was discovered in 1859 near the town of Titusville in Pennsylvania, where a prospector named Edwin Drake drilled into the ground with a derrick, steam engine, rope and bit – a technique already in use for drilling salt wells. At 21 metres below the surface, his well struck oil. Drake never made a fortune from his find but the value of liquid petroleum was quickly realized and soon prospectors were drilling all over America in search of the new fuel. Drake's discovery had marked the birth of the modern petroleum industry, which now provides nearly half of the world's energy.

Today companies searching for oil use many methods to investigate possible sites before drilling even begins. Underground rock formations are carefully surveyed to determine the best place to explore. Surveyors may first seek out variations in the Earth's natural magnetism which can give a clue to the type and thickness of rock that will be found below the surface. Underground rocks containing iron cause a distortion in the Earth's magnetic pull, which can be recorded on very sensitive instruments, but the influence of these rocks deep in the Earth's crust will be lessened if they are covered by thick layers of non-magnetic, oil-bearing rocks. Possible oilfields can therefore be detected by measuring the distortions in magnetic pull caused by the rock pattern, a job usually carried out by aircraft equipped with devices which can record magnetic variations over large areas in a short time.

Magnetic surveys can give only an approximate guide to underground rock conditions. A more detailed picture can be obtained by setting off a small explosion at or just below ground level and recording the time the shock waves take to bounce off the deep rock layers and return to a series of sensing devices strung out along the ground at regular intervals. This method is known as seismic surveying. The speed at which the shock waves travel back varies according to the type and thickness of the underlying rocks. The echoes that return to the surface are recorded by the sensing devices and the information is fed into a computer to produce a picture, known as a seismograph, which indicates possible oil traps.

No matter how much information these geological surveys provide, it is impossible to predict with certainty that oil will be present at any particular place where the rock conditions look right. The facts will only be known after a drilling rig has been set up and a well has been sunk – and that can be a long and expensive business.

Left: potential oilfields can be mapped by sending shock waves, or seismic signals, into the ground and measuring the time the echoes take to return to an array of microphones. Below: a print out of seismic signals, coloured in by geologists to show potential oil-bearing formations.

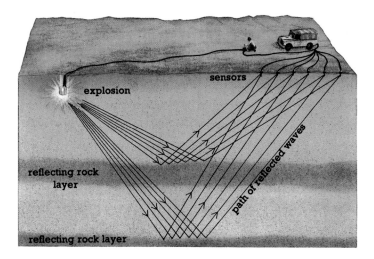

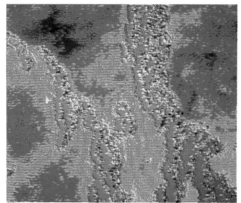

Infrared satellite pictures provide valuable clues for oil prospectors. In this computer-enhanced view of part of the northern Sahara the blue areas represent dome-like structures within which oil may collect. Sand dunes show as red and gravel plains as green.

'Gushers' like this one in Iran in 1917 used to be the first dramatic sign of an underground oil reservoir in the early days of prospecting. Today's technology helps prevent such waste when oil is struck.

Drilling techniques have advanced a long way since the early days of oil prospecting. Drake's drill worked like a pile driver, hammering a hole into the rocks, but modern drills employ a circular motion to do a much neater and faster job. The first rotary oilfield drilling was tried out in Texas during the 1890s and has since become almost universally adopted. It has led to the design of a completely different, more sophisticated type of bit using teeth made from diamonds or tungsten carbide which give the hardness necessary to crush and chip any kind of rock likely to be met with in drilling operations. With this technique, it is possible to drill through more than 300 metres in a few hours, compared with a rate of just 20 metres a day for even the fastest percussion drilling.

During drilling prospectors keep careful records which show the type of rock the bit is passing through. This can be done by retrieving rock samples, but modern techniques also allow drilling crews to decide whether a formation is likely to contain oil at a much earlier stage. Electric currents sent through the ground from the surface are received by instruments lowered into the well at the end of a wire. Because rock containing water is a much better conductor of electricity than rocks where the pores are clogged by oil, variations in the strength of current reaching the instruments give a clue to whether oil is present. Explorers sometimes also use sound signals or even nuclear radiation to help them discover the contents of deeply buried rocks.

Producing Oil

The early prospectors usually got no clear hint of success until they heard a brief rumbling noise and a column of oil as hot as 80 degrees Centigrade suddenly shot out from the well. However, "gushers" were also a terrible waste of oil. Thousands of gallons of oil, which had been sealed under high pressure in a subterranean trap for millions of years, would shower down on the surrounding landscape while the drillers celebrated.

In contrast, today's oilwells are designed with the goal of never spilling a drop. Not only are measures taken to prevent oil from gushing from the top of the well: modern technology can also guard against any being lost underground on its way to the surface – for example, by escaping through cracks in the rock. This is done by lining the well with steel tubing called casing which is firmly cemented into place as the drilling progresses. Besides helping to prevent the sides of the well from caving in, the casing allows a chemical mixture known as drilling mud to be circulated during the drilling operation. The fluid is pumped down through the column supporting the drilling bit and comes out through holes on the bit to help cool the cutting teeth and wash away broken

Bits have now been developed to cope with every conceivable sort of rock. Shown here are seven types fitted with diamond or tungsten-carbide teeth. specially adapted to cutting through hard strata.

As a drilling bit grinds through rock, a clay and water mix known as drilling mud is injected down through the core of the drill pipe to surround the bit and keep it cool. The mud then carries rock particles back up to the surface between the drill pipe and the casing.

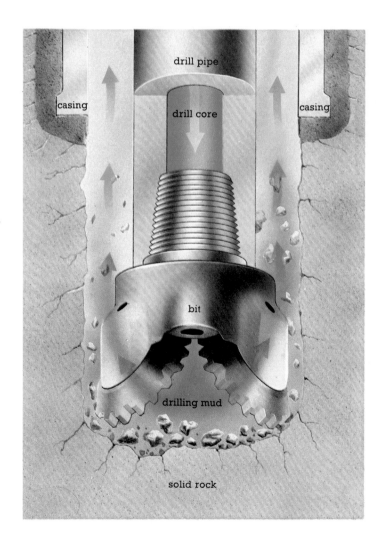

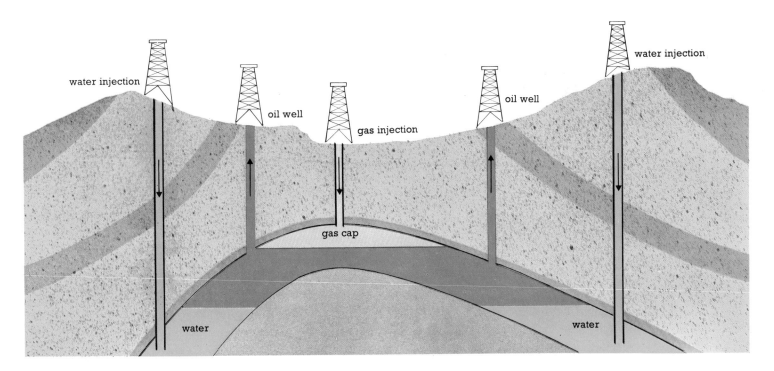

water injection

oil well

gas injection

water injection

oil well

gas cap

water

water

rock. As pumps on the surface continue pushing down more mud, the used fluid returns through the casing to the surface where the drilling crew can usually tell what kind of rock they are penetrating by examining the particles in the mud.

The drilling mud also provides a heavy column of liquid to counteract any sudden build-up of pressure at the bottom of the well, as would happen if the bit pierced through the rock above an oil reservoir. The oil is also held in check by fitting a device called a blow-out preventer to the wellhead. This contains a series of valves which will be triggered off by a sudden increase in pressure from below, sealing off the well as a second line of defence.

When oil is struck another set of valves will be fitted to regulate the flow. This device, which in some big oilfields will be taller than a man, is called a Christmas tree because it resembles a trunk with branches. Most large oilfields will have a number of wells, each fitted with a Christmas tree and connected by pipeline to tanks where oil will be stored.

The natural pressure which causes oil to gush to the surface will normally keep an oilfield flowing for months or even years after the first well has been drilled. But eventually the pressure will drop to a level where it is no longer sufficient to produce a flow, and at this stage up to three-quarters of the total volume of oil may remain in the rocks. When this happens the flow of oil can be boosted by reproducing the natural pressure of the water and gas

The layout of a typical oilfield, showing production wells drilled into oil deposits and shafts to boost the flow of oil drilled into gas caps and water-bearing strata. Because the oil reservoir is not always one large connected pool more than one oil well must be sunk.

between which it is sandwiched. Extra wells can be drilled to inject water into the zone below the oil, or gas into the area above. The gas can be collected at the surface after bubbling out with the oil and will be compressed to boost its pressure before being returned to the natural underground reservoir. Water injection, however, demands a ready supply of pure water which may be difficult to obtain at oilfields in the desert or other arid zones.

In spite of all the efforts made to regulate the flow of oil from a well, natural pressure will sometimes cause an uncontrollable blow-out. The flammable mixture of oil and gas may even catch fire if there is a spark about when it reaches air at the surface. When this happens, oil companies call in a specially trained team who set off explosives amid the flames to expel the air that helps to support the burning. The team then works under the downpour of oil to 'kill' the blow-out by installing valves through which the flow of oil can be controlled.

Oil from the Oceans

In parts of the world where oil was discovered near coastlines, explorers often found that the deposits extended far out beneath the sea. At first offshore oilfields were reached by building piers as far out as possible to support drilling equipment. However, it soon became clear that other techniques would be needed to drill for oil lying under deeper water such as that off the Californian coast.

During the last few decades new techniques have been developed to tackle oil exploration at sea. One of the earliest was the jack-up rig. It gets its name from a set of steel legs which rest on the seabed and can be extended to jack up the drilling platform above the reach of the waves. The legs can also be raised through the deck so that the platform can be towed to another location. Jack-up rigs can operate in water depths of about 90 metres.

Another drilling rig has been specially designed for exploration in deeper water. This has huge

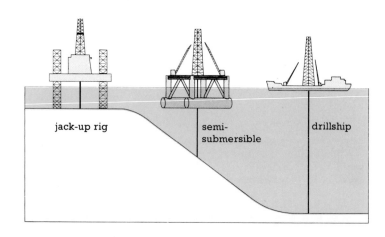

Jack-up rigs are suited to relatively shallow waters with a firm sea bottom; semi-submersibles can operate at greater depths and are stabilized by pontoon floats submerged about 18–30 metres below the surface; drillships, though the least stable, can operate in almost unlimited depths and drillings have been made in water nearly six kilometres deep.

buoyancy tanks which enable it to float out to the drilling site. There the tanks are partly filled with water, making the rig sink lower in the sea and giving it more stability in stormy weather. Anchors are also used to keep it in place above the well. This type of rig, known as a semi-submersible rig, allows exploration in water depths of more than 300 metres.

For exploration in even deeper water, oil companies use a specially equipped drillship. In addition to the usual rear-mounted propeller, this type of vessel has thruster units installed in the hull, enabling it to move in any horizontal direction. These prevent the ship from being moved out of position by tides and currents as it drills. The action of the thrusters is controlled by a computer installed on board. Signals from a beacon fixed on the seabed warn the computer as soon as the ship starts to change position. The computer then turns on one or more thrusters to produce a force which counteracts the movement and keeps the vessel directly above the oilwell.

When an offshore well starts producing oil, a more

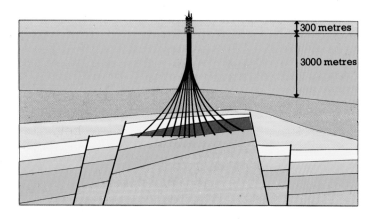

Above: directional drilling enables more than 20 borings to be made from a single platform. The drill is deflected from the vertical at a rate of 2–2½ degrees per 30 metres by wedges placed within the drill hole, and is guided to its destination by sophisticated instruments based on the magnetic compass. The drill hole may be run through the producing layer and anchored in the rock beneath. Oil then flows into the pipe through holes perforated in the casing by steel bullets or an explosive charge.

A jack-up drilling rig being towed into position off the coast of Dubai on the Persian Gulf. The platform jutting out on the left provides landing space for helicopters.

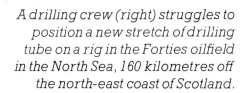

A drilling crew (right) struggles to position a new stretch of drilling tube on a rig in the Forties oilfield in the North Sea, 160 kilometres off the north-east coast of Scotland.

permanent structure is needed to house the production equipment. Usually a huge steel or concrete platform is built and floated out to the oilfield. The base is lowered to the seabed by flooding ballast tanks with water. Meanwhile, strong legs or columns are fitted to the base in sections until they are tall enough to support the production equipment above the surface waves. Some of the biggest offshore platforms have been installed in the North Sea between Britain and Norway, an area which has been found to contain some of the world's richest offshore deposits. Sometimes pipelines carry the oil hundreds of kilometres to the nearest land, but in other cases the pipeline will lead to a nearby buoy where the oil can be loaded on board tankers for shipment ashore.

Oilfields that lie beneath the sea may cover an area of some 75 square kilometres – the size of a modern city – and a large number of wells need to be driven into the oil reservoir to drain all parts of it. Because production platforms are so expensive to set up, it would be too costly to place one above every well, so a cluster of wells is drilled from one platform, fanning out over a radius of 1.5 kilometres or more.

Despite these developments, drilling platforms may one day seem as old-fashioned as the hand-operated rigs of the first prospectors. Wellheads could be constructed on the seabed and maintained by engineers working in small submarines. The oil would be piped to a surface mooring buoy where it could then be loaded on board tankers.

Oil from Sand and Shale

Oil deposits are not always buried deep down in the Earth's crust. In some places oil has been formed under just a thin layer of soil and rock. This layer has been too light to compress the oil-bearing sands into the firm rock typical of deep underground oil reservoirs. Even so, there has been sufficient heat at these shallow depths in the Earth's crust for hydrogen and carbon from decomposing plant and marine life to combine over millions of years into crude oil similar to that found at much greater depths.

Huge amounts of oil in this form are found along the Athabasca River in Alberta, Canada, where the deposits – known as oil sands – are near enough to the surface to be removed by a method similar to strip mining. This field is part of a much larger area of oil-bearing sands in Alberta that probably contain more oil than all the oilfields of the Middle East, but only ten per cent is suitable for surface extraction.

The recovery technique involves first removing as much as 45 metres of overburden to allow huge mechanical diggers to start scooping out the black, oily deposits. The operation is very complex and so very costly, and oil companies use huge digging machines able to handle more than 6000 tonnes of material an hour. Once excavated the oil has to be separated from the grains of quartz that it coats. Unlike other oil sand deposits – for example, those in the United States – this oil is not in immediate contact with the sand but is separated from it by an 'envelope' of water which surrounds each grain. This means that the oil sand can be processed by mixing it with water and steam, producing a froth which is spun at high speed in huge tanks. The oil then separates from the sand in the same way that water is separated from wet garments in a spin-dryer.

oil sands

oil shales

This map shows the areas of oil sand and shale in the United States and Canada.

In the future it may be possible to produce this type of oil without costly excavation and processing. By drilling wells into the oil sand layer, engineers could inject hot water and steam through some of the wells to create heat and pressure that would break down the forces causing the oil to stick to the sand. This should form a free-flowing liquid which could then be collected through other wells, in much the same way as crude oil flows to the surface from deep-lying reservoirs in other parts of the world.

Another type of oil that can be recovered by mining rather than drilling is found in greyish rocks known as shales. The oil is chemically combined with the shales, and very high temperatures are needed to force the rock to decompose so that it will give off its oil content. Although this makes shale oil very costly to extract, in the future it could become an increasingly important energy source. It has been estimated that there is much more oil locked up in shales than in all the world's conventional oilfields.

Some of the richest shale deposits are found in the states of Colorado, Utah, and Wyoming, where it was discovered by pioneers in the early nineteenth century. However, the world's first commercial use of shale oil was based on mining operations in the Lothian region of Scotland in the middle of the nine-

One of the huge draglines that remove oil sands from the main excavation areas along the Athabasca River in Alberta, Canada, where two commercial plants are capable of producing 150 000 barrels of oil per day.

teenth century, after a scientist, known as James "Paraffin" Young, devised a process of removing the oil by roasting the rock in ovens and draining away the fluid that it gave off. Extraction techniques have been much improved since then, but even so, about a tonne or more of waste rock will be produced for every 350 litres of oil recovered.

The mountains of waste rock created by shale processing are a serious drawback, but now ways are being found of extracting shale oil without having to take the rock out of the ground. For example, by generating heat in the rock formation the oil can be made to flow freely enough to be drained off by pipelines or wells. Controlled underground explosions might also be used in the future to fracture shale rock formations and allow a mixture of steam and air to be piped through the cracks, turning the oil into a vapour which could be collected by pipeline. Exploiting shale by this type of process could provide the United States with oil that compares favourably in cost with supplies from the Middle East.

A polished slab of Green River oil shale from the Rocky Mountains. The dark brown streaks are organic material which will yield oil when heated.

Arctic Oil

The discovery of vast oilfields beneath the frozen waste of Alaska confronted technicians and engineers with a spectacular challenge. Oil could hardly have been found in a more remote and inhospitable place. The supplies are thousands of kilometres from the industrial centres where fuel is needed, and because the Arctic seas are frozen for much of the year, it is impossible to carry cargoes of oil by sea.

When the first big oilfield was discovered at Prudhoe Bay on the northern coast of Alaska in 1968, oil companies decided to construct an overland pipeline to carry the oil many miles south to the ice-free port of Valdez, where tankers could collect supplies throughout the year. Constructing this pipeline was one of the most ambitious feats of civil engineering ever attempted. The line had to cross hundreds of rivers and streams, scale mountain passes nearly 1.5 kilometres high and cross many kilometres of desolate terrain that forms one of the world's biggest tracts of natural wilderness. People and materials had to be moved into remote locations to build eight pumping stations that would boost the flow of oil on its week-long journey south.

The greatest challenge, however, was posed by the delicate ecological balance of the tundra, where

Because oil is warm when it leaves the ground and is continually reheated by the compression pumps that keep it on the move, much of the Alaska pipeline is supported on stilts. These prevent the warmth of the oil melting frozen layers of ground and causing permanent ecological damage.

ground temperatures rarely rise above the freezing. Over vast stretches this has resulted in a layer of up to 600 metres of frozen subsoil, known as permafrost, which lies just below the surface and never melts. If it ever did thaw out and release its water content, it would turn great areas into swamps, and cause irreparable damage to rare species of wildlife and Arctic plants which have adapted to the demanding environment. Oil companies were therefore not always able to follow their normal practice of burying the pipe with its cargo of warm oil.

The problem was overcome by raising the pipeline on stilted supports for much of its route. These supports also leave room for the line to bend from side to side as the metal contracts or expands with the heat of the oil each time the flow stops and starts. At places the pipeline has even been raised to allow herds of migrating caribou to pass underneath while at other places the supports have been placed farther apart to allow the line to move sideways or upward without breaking during one of the notorious Alaskan earthquakes.

This map shows the route of the Alaska pipeline, built across 1300 kilometres of rugged terrain to carry oil from Prudhoe Bay on the edge of the Arctic Ocean to the port of Valdez in southern Alaska.

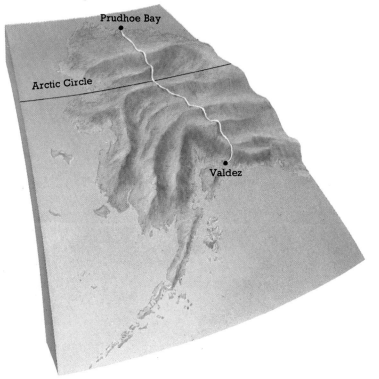

An artificial island, built in the Beaufort Sea off the coast of northern Canada, serves as a base for exploration but may later be used as a permanent production facility.

The discovery of oilfields on the edge of the Arctic Ocean has encouraged oil companies to look for offshore oil deposits in the same area. Because pack ice forms on the surface of the sea for most of the year, exploration with conventional offshore drilling rigs is impossible: their tubular steel legs would be crushed by the pressure of the ice. Oil companies have solved this problem by building artificial islands in the comparatively shallow water during the short ice-free period in the summer. These islands, made of gravel dredged from the surrounding seabed, are able to support the weight of a normal land-based drilling unit which can be transported in sections across the surface of the frozen ocean in winter.

Refining Oil

Crude oil is a highly complex mixture of many different combinations of hydrogen and carbon with traces of other elements. The actual composition varies from one oilfield to another and one way in which chemists test the value of a particular type of crude oil is by finding out how much sulphur it contains. A high proportion of sulphur causes atmospheric pollution if the oil is burned as fuel, and crude oil with a high sulphur content is described as 'sour' because of its unpleasant smell. A lower sulphur content produces a 'sweet' oil which is better suited to producing fuels for automobiles and heating systems.

In order to provide fuels, crude oil has to be refined. The liquid is first boiled to turn it into a light vapour which is passed into a tower where it rises and cools. Each of the components of crude oil has a different boiling point, so at each stage of the cooling a different liquid, or fraction, can be drawn off. Parts of the mixture that are not fully vapourized at first will be collected as a liquid at the bottom of the tower and recirculated.

At the end of this stage, known as fractional distillation, the crude oil mixture will have been separated into several distinct substances. The components which boiled at the lowest temperature will include propane and butane. Liquids which boiled at higher temperatures will include fuels used in cars and

aircraft while heavier oils needed to power diesel locomotives or to fire industrial boilers would need even more heat. But none of the products will be ready for the consumer until they have been refined even further.

Time and again, products will be vapourized and returned to liquid form as they pass through a series of processes designed to purify them and improve their performance as fuels. For example, most modern refineries reduce the sulphur content in oil products. Besides minimizing pollution, this process yields pure sulphur which can be used in the manufacture of fertilizers. If the first refining stage produces too much heavy fuel oil and not enough of the lighter products such as petrol for cars, the heavier products are passed through another unit known as a cracker which will break them down

Right: this diagram shows the distilling operation of a fractionation column. Crude oil, flowing through pipes, is heated and mixed with steam so that major compounds within the oil rise to varying levels according to their boiling point, enabling them to be drawn off to form a variety of petroleum products. The residue is recycled through the column.

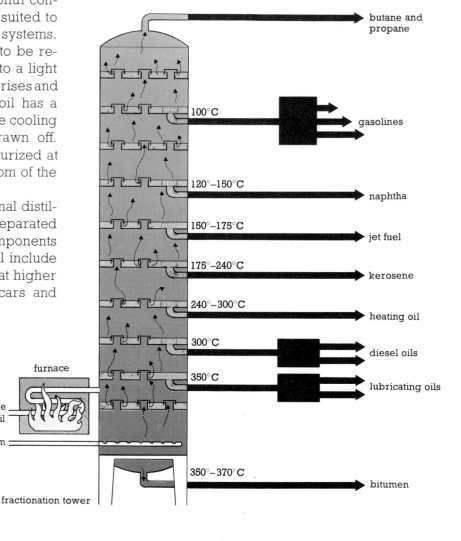

Oil refineries operate around the clock to meet the modern world's continuous demand for fuels and oil-based products. This one, near Southampton, processes 19 million tonnes of oil annually and employs 2200 people. The structure in the foreground removes impurities from gases, while in the background can be seen a catalytic cracker for breaking down heavy oils into lighter ones.

into lighter oils. Finally, the output from various parts of the refinery may be blended and combined with other chemicals to produce a complete range of the fuels most suited to the needs of consumers.

The first oil refineries were designed only to produce fuels. Components in the crude oil that were not needed as fuel were usually just burned off. But over the years, uses have been found for many of these 'waste' products. Nowadays some are turned into lubricants and others are made into plastics. It is even possible to manufacture pharmaceuticals from oil – and these produce far more effective remedies than those once made from natural oil seepages.

Crude oil supplies the material from which an astonishing variety of goods is manufactured. Even the child's plastic bracelet and the wax crayon are made from oil-based substances, and many products we regard as essential to modern life depend on a continuing supply of oil.

Oil in Use

The most familiar use of oil is as a fuel for motor vehicles. Petrol, which is one of the main products derived from naturally occurring crude oil during the refining process, ignites with explosive force when it comes in contact with an electric spark in the cylinders of an internal combustion engine. These explosions take place several hundred times every minute and provide the energy necessary to move the pistons, which provide the mechanical power to drive the vehicle.

However, oil supplies much more than just a source of power for modern vehicles. The refining process also produces a range of lubricants that can be circulated around the moving parts of the engine to minimize friction and keep metal parts cool so that they do not expand and become distorted by heat building up in the engine. Like petrol, lubricating oils are specially blended for the job they have to do. The aim is to produce an oil thin enough to create no resistance to moving parts when cold but thick enough to remain effective when the engine is running at operating temperature.

In most designs of car, engine cooling is provided by circulating water to pick up some of the heat, and by a fan that supplies a constant stream of cool air. Both the hoses through which the cooling water passes and the fan may be made from petrochemical materials derived from crude oil, and many of the other materials in the vehicle will also be made from oil-based products – for example the foam forming the seats, the dashboard, the roof covering, the carpeting and the steering wheel. The tyres are yet another petrochemical product, and even the road they travel along may be made of the oil-based substance bitumen.

Oil products are, in fact, an indispensible part of our modern lifestyle. The complex array of hydrocarbons contained in crude oil can be processed to provide everything from medicine and fertilizers to plastic substitutes for paper, glass and metal. Although many of these things could also be made from coal, oil has the advantage of not having as high a content of the heavy carbon element, which means that it is much lighter to transport. Also, since it occurs naturally as a liquid it is much cheaper to process than coal.

In spite of the wide range of materials that can be produced from oil, much of the world's supply has in the past been burned as a fuel providing heat in homes, factories and power stations. Because crude oil could in the future be in short supply, this is now considered a wasteful use of a prime resource. Many scientists believe that there should be restrictions on the burning of oil to provide heat which could just as easily come from coal or nuclear power. They argue that the world's remaining oil supplies should be reserved for the purpose to which they are best suited – the provision of high-quality transportation fuels for cars and aircraft and as raw material for petrochemical processing.

Rolls of fabric made from oil-based polymer fibre, one of today's most widely used textile yarns, awaiting dispatch to a clothing manufacturer.

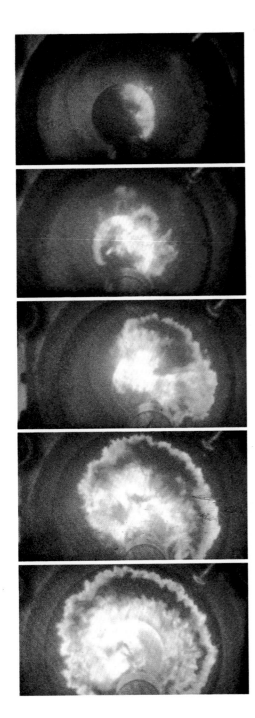

This sequence of photographs, taken through a quartz panel, shows the combustion of diesel fuel within the cylinder of an internal combustion engine, a process which lasts no more than three hundredths of a second.

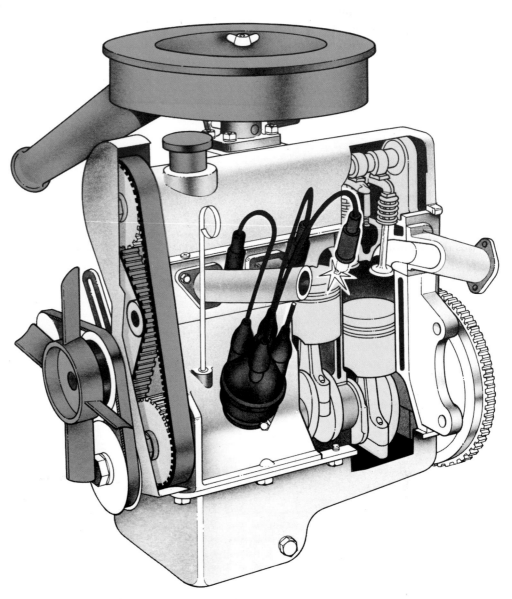

Above: an internal combustion engine depends on oil for more than just fuel. This diagram of a typical car engine shows parts made from oil-based products (green) and the area bathed in lubricating oil (orange) to ensure smooth running.

Gas from Coal and Oil

Toward the end of the eighteenth century researchers found that heated coal produced a brightly burning gas. This discovery was quickly put to use by factory owners who used the gas to light their dingy workshops during the early days of the Industrial Revolution. Later, coal gas was widely used to light streets in the expanding towns and cities, and its popularity increased still more after Carl Auer von Weisbach, a scientist working in Germany in the 1880s, discovered that eight times as much light could be produced by burning the gas through a cloth mantle treated with chemicals that glowed brightly when heated.

The use of gas for street lighting led to the building of gas-making plants, or gasworks, in most towns and cities. After pipes had been laid underground to provide lighting in the streets, it was a small step to start piping gas into homes, not just for lighting but also to provide heat for cooking and comfort. Gas was a much more convenient household fuel than coal, which had to be piled regularly on the fire and produced ash that had to be removed later. In effect, piped gas meant that all the stoking and cleaning out of fires was carried out by someone else at the town's central gasworks.

Through the nineteenth century and the first half of the twentieth century, gas was manufactured in towns and cities throughout Europe and in many parts of North America by loading coal into enclosed chambers known as retorts. These were simply big ovens in which the coal could be heated in the absence of oxygen – which would have set the coal alight and burned up the gas. Temperatures of more than 950 degrees Centigrade inside the retorts produced a mixture of gases from the hydrogen and carbon

A worker rakes flaming coke from a gas-making retort at one of the last coal-gas plants in Scotland during 1979 – its last year of operation. Gas from coal burning in the retorts rose through pipes, visible in the centre of the picture, and was distributed to 300 households.

contained in the coal. Part of this mixture was pure hydrogen, part was carbon monoxide, made up of carbon and oxygen, and the rest was mainly methane – a combination of hydrogen and carbon. As the gases were produced, they were piped out of the retorts and treated to remove unwanted impurities such as sulphur and nitrogen, and then delivered to customers. The process also produced a number of by-products including ammonia for manufacturing chemicals and coal tar which could be used for a range of products like drugs and artificial dyes. And the coal from which the gases had been removed came out of the retorts as coke which could be used in steel making.

During the 1950s a process for producing gas from oil began to replace the traditional method of manufacturing coal gas. At that time, there was a shortage in Europe of the coals best suited to gas manufacture but an abundance of cheap oil from the fields in the Middle East. The new process used

A modern coal gasification plant at Dörsten, West Germany, where gas is manufactured by burning coal in a high-pressure mixture of oxygen and steam.

naphtha, one of the products of oil refining, to produce a mixture of gases similar to that obtained from coal. This is done by mixing naphtha with steam in an enclosed chamber containing a chemical that acts as a catalyst. The catalyst aids a chemical reaction that produces a mixture of carbon monoxide, hydrogen, and methane.

Nowadays electricity has taken over from gas as the principal source of lighting in streets and buildings but gas still provides a convenient fuel for cooking and heating in many households as well as being used in industry. Gas produced from coal or oil can now be passed through additional stages which boost the methane content. Even if naphtha becomes difficult to obtain because of worldwide shortages of oil, it should be possible to use a wide range of coals to produce gas that has twice the heat potential of the old form of gas in each cubic metre.

Natural Gas

In many places gas is formed naturally in the Earth's crust. At least 2000 years ago the Chinese boiled pans of salt over a flame produced by burning natural gas brought from great depths through bamboo pipes. The Ancient Greeks, too, discovered that seepages of what they called 'the breath of Apollo' caused sheep to become dizzy in nearby fields. We now know that natural gas is often produced when coal seams are heated by high temperatures deep underground – indeed, the lack of oxygen among the buried layers of rock helps to make the Earth's crust a perfect gas-manufacturing oven.

Much natural gas, however, occurs independently, rising toward the Earth's surface and building up in pockets if its progress is blocked by an upper layer of solid rock. The gas accumulation can be reached by drilling wells in the same way as oil is recovered from an oilfield. Natural gasfields of this type provide a valuable source of energy in the United States, which produces 35 per cent of the world's supply. Similar discoveries in Western Europe, Canada and the Soviet Union during the last 20 years have led some countries to abandon the manufacture of gas from coal and oil.

Oilfields also contain natural gas, which builds up at high pressure above the oil. The oilfields of the Middle East, for example, once contained about a fifth of the world's natural gas reserves. In the past huge quantities of gas recovered along with oil were burned as waste because there was no use for the energy locally, but with other parts of the world running short of fuel, the gas is now retrieved as a valuable source of energy. Although at present rates of consumption the world could exhaust its

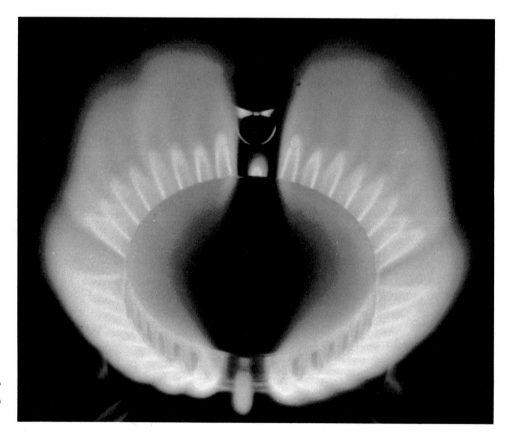

The clean, easily controlled flame of natural gas is ideally suited to domestic appliances.

Natural gas being flared as waste in a Middle East oilfield. Scenes such as this are becoming rarer as increasing amounts of gas are liquefied for transport to energy-hungry industrial nations.

known reserves of natural gas within about 50 years, new fields are being discovered yearly and natural gas may provide us with a valuable energy source well into the next century.

The composition of natural gas varies depending upon its origin. Most gas contains a high proportion of methane, a clean-burning and odourless fuel which can give twice as much heat as the same quantity of some manufactured gases. Methane is widely used for domestic and industrial heating. Oilfields, on the other hand, produce a gas with other combinations of carbon and hydrogen which become liquid under the slightest pressure. The components in this gas include propane and butane which can be bottled to produce cooking and heating gas.

All natural gas is dried by passing it through chemicals that absorb moisture. Acid impurities – for example, hydrogen sulphide and carbon dioxide, which have been formed along with the gas – are removed by bubbling the gas through a liquid that absorbs acid. Distribution companies also frequently add chemicals which give the usually odourless substance a nasty smell so that customers will be able to detect leaks instantly.

Pipelines and Supertankers

Oil and natural gas can be transported in their natural states through pipelines buried underground or even on the seabed. Hundreds of kilometres of underground pipeline have been laid to transport oil and gas ashore from important offshore producing areas such as the North Sea and the Gulf of Mexico. There are also land pipelines carrying both oil and gas products over thousands of kilometres between producing areas and centres of population in North America, as well as in the Middle East.

Some of the earliest pipelines used to move fuel were made from wood. In the United States during the nineteenth century, holes were drilled through the centre of tree trunks to provide tubes for gas distribution with inside diameters of more than seven centimetres. Nowadays main distribution lines for crude oil and gas are almost always made of steel. The largest one, crossing Alaska from Prudhoe Bay to Valdez, has a diameter of 1.2 metres and would be able to carry 100 million tonnes of oil a year if used to its full capacity.

Most land pipelines are buried at least one metre underground, often under fields where crops are growing undisturbed by the fuel passing beneath. Underwater pipelines may also be buried in a trench on the seabed. All oil and gas pipelines are coated with a layer of bitumen or fibreglass to prevent corrosion while underwater pipelines have an additional

A section of the new European trunk gas pipeline ready for laying beneath the Main River in West Germany. This dual pipeline will transport about 10 billion cubic metres of gas per year from the Soviet Union to West Germany with branch lines taking supplies to France and Austria.

In this liquefied natural gas tanker over 125 000 cubic metres of gas is carried at −178° Centigrade in five aluminum spheres insulated with polyurethane foam and fibreglass. The vessel carries about half as much energy as an oil tanker of equivalent size.

The Esso Dalriada, a modern supertanker, has a length of 340 metres and can carry up to 266 000 tonnes of crude oil.

coating of concrete for extra protection against the effects of seawater. Long pipelines usually need pumping stations every 100–250 kilometres to give the oil or gas an additional boost along the way, and in remote areas these stations sometimes get their energy supply by using a minute quantity of the fuel being carried in the line.

Although pipelines are a very efficient way of transporting oil and gas, they need to be regularly cleaned to remove wax and other deposits which the fuels leave on the inside walls, particularly where there are bends in the line. Cleaning is carried out by using the oil or gas flow to push through a device known as a 'pig'. This odd name comes from the initials of 'pipeline inspection gadget', which describes another of its uses. The pig has a diameter which exactly fits the inside dimensions of the pipe and will scrape away blockages with its outer edge as it moves along. However, if the pipeline is dented, the pig will have to be pushed back by pumping through high-pressure air or water in the other direction while repair crews are called out to replace the faulty section of line.

Much of the world's oil comes from areas, such as the Middle East, that are too distant from the main markets to make transport by pipeline either economic or practicable. Oil from these regions is shipped to North America, Europe and Japan in specially built tankers. The first oil tanker, the Glückauf, launched in 1866 could carry just 300 tonnes of oil but modern vessels can carry half a million tonnes. These supertankers are more than 400 metres long and hide their bulk beneath the surface like icebergs. When fully laden, such ships require a draught of more than 30 metres and their routes avoid areas where the ocean is shallow.

In the future increasing quantities of natural gas will also be transported by sea from the United States, the Middle East and other areas to the industrialized nations. This is a difficult operation because natural gas, at normal temperatures, takes up a thousand times as much space as an amount of heavy fuel oil yielding the same amount of heat. But the problem can be overcome by turning the gas into a liquid which reduces its volume by 600 times. The liquefied gas can then be loaded onto insulated tankers. Even in liquid form natural gas is lighter than oil, so the tankers sit much higher in the water than crude oil carriers, but they can be two or three times more expensive to build for each tonne of cargo carried. At unloading terminals, the cargo is heated and pressurized until it turns back into gas and is then distributed through high-pressure pipelines.

The Nature of Electricity

The effects of electricity could be seen long before anyone realized that it would become one of our most useful forms of energy. Lightning flashes during a thunderstorm are the most dramatic form of electrical activity found in nature. But the Greeks discovered a more down-to-earth way of creating electricity. By rubbing a cloth with a natural resin known as amber, they were able to produce a strange force that caused bits of dust to stick to the resin. They called the mysterious force 'elektron' which is simply Greek for amber. Modern science has adapted this word – as electron – to describe the tiny particles in atoms which we now know cause this phenomenon.

Normally electrons circulate in the outer part of atoms, held in position by an invisible bond, known as an electrostatic force. This bond is created by the attraction of the electrons, which are negatively charged, to the positively charged protons in the core of the atom. Sometimes, however, the electrostatic force can be overcome, allowing the electrons to break free from their orbits. This can happen when heat is generated by friction as two different surfaces are vigorously rubbed together. In the method discovered by the Greeks, electrons freed from the surface of the cloth became attached to the surface of the amber. The amber already has electrons of its own, so the electrons from the cloth are surplus to requirements and create an additional negative charge on the surface of the amber. At the same time, other spare electrons attach themselves to very small specks of dust, light enough to be attracted by the electrostatic force. The amber would stay charged with this energy, known as static electricity, until brought into contact with another material which would readily absorb the spare electrons, such as the cloth from which they had broken loose during the rubbing.

Static electricity in much more dramatic form causes the lightning which lights up the sky during thunderstorms. Interaction between raindrops or ice crystals leads to enormous build-ups of electrons in the lower regions of a cloud. The only way in which the balance of protons and electrons can be restored is for some of the electrons to leap to the ground or to a positively charged part of the cloud, producing a flash of light and a subsequent boom of thunder. Lightning can cause great damage and even kill people by the intense heat generated from the electrostatic charge. But understanding of electricity has helped to safeguard man-made structures from this natural hazard by the use of metal conductors installed at the tops of buildings. These collect the huge accumulation of electrons and dispatch them safely into the ground. The charge will then be neutralized as the electrons attach themselves to the abundant atoms in the Earth's crust.

Metals are good conductors of electric current because of their structure. Their atoms are packed tightly together in a lattice-like pattern, so tightly in fact, that the electrons are squeezed. Some of the

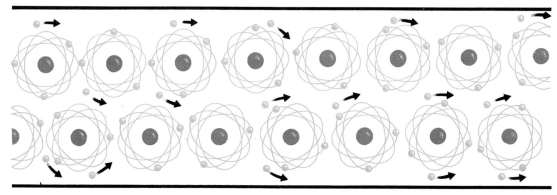

This schematic drawing shows two rows of atoms in a metal conductor, such as copper wire, and the flow of electric current when free electrons move between atoms in one direction.

Lightning flashes such as these crackling over Johannesburg represent a concentrated burst of energy that generates temperatures of up to 15 000° Centigrade.

electrons farthest away from the protons consequently become dislodged from their normal paths. The result is that these electrons are free to move through the lattice. When they all move in one direction an electric current flows through the metal. The movement of electrons takes place very quickly, which is why electricity seems to pass instantly from one place to another.

The electrons of many other substances are held firmly in orbit by their protons. This means that the outer electrons are less likely to break free. Such materials, which are poor conductors of electricity, include rubber, plastics, wood, and the amber which the Greeks used in their early experiments. These substances are known as insulators and can be used to stop the passage of an electric current. That is why electrical conductors are usually coated with an insulating material such as plastic.

Batteries and Bulbs

Centuries after the Greeks discovered the electrical properties of amber, scientists found that certain chemical reactions could also generate electricity. For example, if a wire is attached between rods of zinc and copper, and both rods are dipped in a container of dilute sulphuric acid, an electric current will flow along the wire. This is because the acid begins to react with the zinc and sets free a stream of electrons. These travel through the acid to unite with atoms of hydrogen known as ions, carrying a positive charge, which are set free from the acid and gather around the copper. The wire between the two pieces of metal provides a path along which there will be a constant

A light bulb turns electric energy into light by passing a current through a thin wire enclosed in a vacuum or surrounded by an inert gas. Although the wire becomes white hot, the absence of air prevents it from combining with oxygen and burning.

flow of electricity until the reaction is complete.

This is just one of many combinations of acid and metal that can be used to produce a battery converting stored chemical energy into electricity. The time a battery will continue to generate electricity depends on the type and on amount of material used and the amount of energy being drawn out. For example the acid need not necessarily be in liquid form. It can be made as a moist paste to minimize the risk of leakages, as in the case of so-called dry-cell batteries manufactured for flashlights and transistor radios. Dry-cell batteries usually have to be thrown away and replaced when their power runs out but scientists have also learned how to make rechargeable batteries in which the chemical reaction can be reversed to take in an electrical charge and then be switched back to provide power. Cars have this kind of battery, which is charged by an electric generator

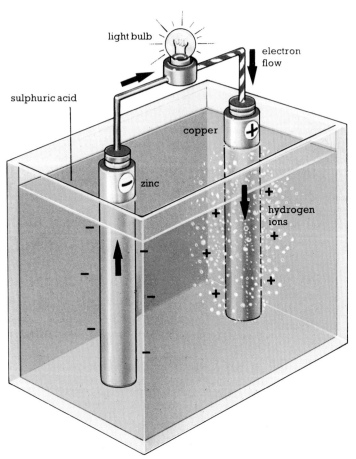

light bulb

electron flow

sulphuric acid

copper

zinc

hydrogen ions

When zinc and copper rods are immersed in dilute sulphuric acid, the zinc sets free a stream of electrons that cause an electric charge to pass between the rods and along the wire connecting the two metals. These electrons unite with hydrogen ions around the copper to create hydrogen gas. This is the principle on which electric batteries work.

while the engine is running.

Batteries can also be used to power road vehicles directly. Because the most efficient batteries contain a lot of lead to react with the acid in their cells, they tend to be very heavy and this creates extra weight for the vehicle to carry. It can also take 80 hours of recharging from a domestic electricity supply for a battery-driven vehicle to pick up as much power as a petrol-driven car would receive in a minute at the petrol pumps. However, research is now taking place into the development of lighter and more efficient batteries which will extend both the range and power of electrical vehicles. In the future it might even be possible for filling stations to stock recharged battery packs for exchange by motorists who are running out of power on long journeys.

An electric current produced in a battery can be converted into light by passing it through a thin filament. Scientists attempted to produce a practical electric light throughout the later part of the nineteenth century. In 1878 the British scientist Sir Joseph Swan showed that by treating a cotton thread with sulphuric acid and placing it inside a glass bulb from which air had been excluded a bright light could be produced. This discovery was the forerunner of the first fully practical electric light bulb, developed independently by the American inventor Thomas A. Edison. On 19 October 1879 Edison lit a lamp containing a filament of carbonized sewing thread, which continued to burn steadily for two days. Further experiments produced a lamp with a life of several hundred hours and in December 1879 Edison demonstrated his invention to an amazed public. The commercial possibilities of electric lighting were quickly realized and within the next three years over 150 of Edison's lighting systems were installed in places varying from factories to steamships.

Nowadays light bulbs usually use wire made from tungsten, which has a high melting point. They also contain a small quantity of argon, a gas which prevents the tungsten from evaporating in the intense heat. And of course, their power comes not from batteries but from huge power plants that also supply electricity for a wide range of other appliances.

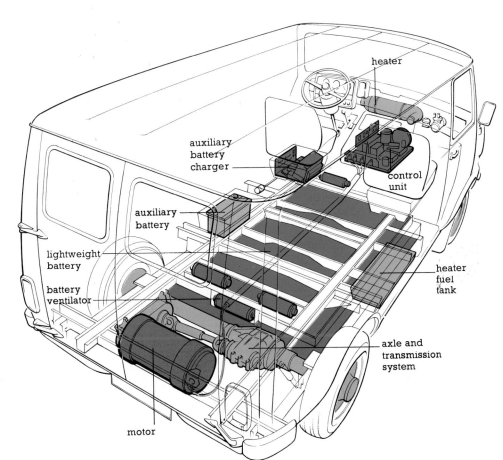

heater

auxiliary
battery
charger

control
unit

auxiliary
battery

lightweight
battery

battery
ventilator

heater
fuel
tank

axle and
transmission
system

motor

This diagram shows the position of batteries, controls and motor in a British-built electric van. All the energy in the main battery is used by the motor, the heating and lights being powered by additional systems. The vehicle has a top speed of 80 kilometres per hour, a range of about 100 kilometres and can carry a load of over 900 kilogrammes.

Electricity for All

In 1820 Hans Christian Oersted, a Danish scientist, discovered an important link between electricity and magnetism. He found that an electric current passing through a wire sets up a magnetic field which can deflect the needle of a compass placed nearby. His discovery was to lead both to the development of the electric motor and to a new way of producing a flow of electric current.

In its simplest form, an electric motor consists of a wire loop positioned between the two poles of a magnet. Each end of the loop terminates in a semi-circular strip of metal, forming a commutator against which are placed two simple contacts known as brushes. An electric current flows into and out of the loop through the brushes, building up a magnetic field. Forces between this magnetic field of the loop and that of the magnet make the loop turn until the two fields point in the same direction. At this stage the motion would stop if the current continued to flow in the same direction but the brushes now make contact with the other halves of the split commutator, causing the current to flow through the loop in the reverse direction. This reverses the magnetic field, causing the loop to make a further half turn and the whole

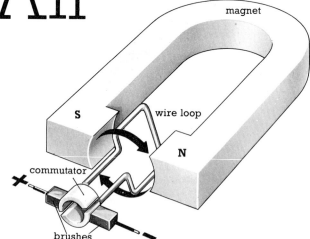

This diagram shows perhaps the simplest form of electric motor. An electric current passes in and out of the loop through brushes which make contact with a commutator formed from two semicircular rings. The current flowing through the loop sets up a magnetic field with north and south poles. Forces of attraction and repulsion between these poles and the north and south poles of the magnet cause the loop to rotate. When the loop rotates half a turn the two halves of the commutator change places, thus reversing the direction of the current. As a result the loop continues to revolve in the same direction.

A selection of domestic electrical appliances reveals the extent to which our modern life-style depends upon a constant supply of cheap electricity.

process repeats as long as a current flows, producing a constant rotary motion in the loop.

About 12 years after Oersted's experiment showed the possibility of combining magnetism and an electric current to create motion, Michael Faraday, an English scientist, showed that a similar machine could produce electricity. He first generated a current by passing a magnet through the centre of a wire coil but the flow lasted only as long as the magnet was moving. A continuous flow was achieved by rotating a copper disc between the poles of a magnet, freeing electrons from the copper and producing an electric current. Faraday's simple generator was the forerunner of machines weighing hundreds of tonnes which now produce enough electrical energy to illuminate several million light bulbs at a time.

A typical modern power station depends on a supply of high-pressure steam produced in coal- or oil-fired boilers, which drives turbines linked to the electricity generators. The turbines use several stages of fixed and rotating blades to give the greatest efficiency in converting the heat to mechanical energy that drives the generators. At the end of its journey the steam is condensed by passing it

The complex generating machinery of a modern power station, such as this one at Drax in the north of England, contrasts strikingly with the tiny magnet motor in the diagram left. The semicircular installation in the centre is the generator, the rotor of which is a 94 tonne electromagnet. Behind it, concealed in this picture, is a series of five steam turbines, flanked by six upright cylindrical condensers which turn steam that has passed through the turbines back into water for recirculation to the boiler. On the far left are two feedheaters, which take steam from various parts of the turbine and use it to heat water flowing from the condenser, thus reducing the amount of energy needed to turn it back into steam.

around pipes containing cold water, and then returned to the boiler where it will be reheated for another cycle through the turbines. Power stations are often built near rivers or the sea which provides a ready source of cooling water. At inland sites where there are no rivers, the water from the cooling pipes is passed through huge cooling towers where its heat is removed by evaporation before it is re-used.

Power from the Atom

Most forms of energy are produced by reactions in which atoms of different kinds combine but remain intact in the process. However, atoms themselves are held together by bonds that will release enormous amounts of energy if they are broken. In 1905 Albert Einstein, a German-born physicist, produced a formula which gave a clue to the amount of energy in these bonds. He expressed this formula $E=mc^2$, with E representing energy, m the change in the mass of the atom when the bonds are broken, and c the velocity of light. This theory meant that the split-

The towering fireball of a nuclear explosion set off on Bikini Atoll in May 1956 dramatically testifies to the huge amount of energy unleashed from the nucleus.

ting of the atoms in one gramme of uranium could give an output of energy equivalent to that from 3 tonnes of coal. It took nearly 30 years before experiments could be devised to prove Einstein's theory, but we now know it is correct. His work had laid the foundations for harnessing atomic power.

The awesome energy released by splitting the atom was demonstrated to the world at large by the devastating effects of the atomic bomb dropped on the Japanese city of Hiroshima in 1945. Such bombs release their vast store of energy in a chain reaction lasting about a millionth of a second, producing a blast that can be as strong as 60 million tonnes of TNT, the most powerful chemical explosive, being ignited simultaneously.

Yet releasing power from the atom is extremely difficult. A way has to be found to break the bonds in the nucleus so that they will yield their binding energy in the form of heat, light and other forms of radiation. However, the normal balance between protons and neutrons can be upset if an atomic particle from outside is brought into a nucleus already overcrowded with particles. Uranium, which is the heaviest naturally occurring element found on Earth, is an example of an atom with an overcrowded nucleus. Not only will some of its atoms readily split if a stray neutron collides with them: sometimes its atoms spontaneously release a neutron, which runs into another uranium atom and causes the same reaction to take place there. This makes uranium an ideal fuel for atomic energy.

There are three different types of atom mixed together in natural uranium ore. One, of which there is only a trace, has 234 particles in its nucleus,

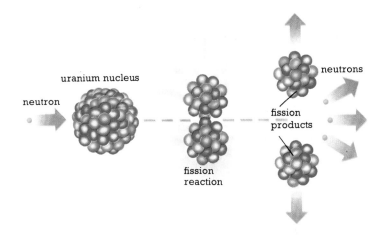

Nuclear fission takes place as free neutrons bombard and split apart the nucleus of a uranium atom, breaking the bonds that normally hold the nucleus together and releasing enormous amounts of energy.

another 235 and the most common type has 238. The atom that has 235 particles, which makes up about seven parts in 1000 of natural uranium, is the most important source of nuclear power. If enough atoms of this type can be brought close together, they will form what is called a critical mass and trigger off a nuclear reaction known as fission. This means that stray neutrons from some atoms will start bombarding others, forcing the particles in the nucleus of the bombarded atoms to split.

When fission takes place, nuclei that are bombarded by stray neutrons do not completely disintegrate, but instead split into two parts, each becoming the nucleus of some other material with fewer particles. The new nuclei will not necessarily be the same size and a fission reaction does not always produce the same new materials every time, but it does set free the huge stores of energy which previously bound a large collection of protons and neutrons together as a nucleus of uranium 235. As the atom splits apart, two or three neutrons will also be released and these will continue the fission process by colliding with uranium 235 atoms which have not yet split. In this way a chain reaction is set up which will continue producing immense amounts of energy until all the uranium 235 in the critical mass has been transformed into a mixture of new materials that do not react with the stray neutrons.

An atom of uranium 238 has 92 electrons arranged in 7 shells. The nucleus contains 146 neutrons and 92 protons, a total of 238 atomic particles.

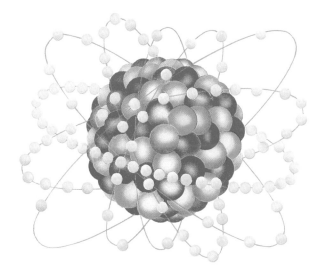

Taming the Nucleus

The atomic bomb releases its energy in an almost instantaneous chain reaction. But if this reaction is slowed down it can be used to provide energy for peaceful purposes. The heat produced will turn water into high-pressure steam that will drive turbines and operate electricity generators. Moreover, much less fuel is needed to produce steam from a nuclear reactor: a tonne of uranium 235 will produce as much energy as about 320 000 tonnes of coal.

Nuclear power stations receive fuel supplies in the form of fuel bundles. These consist of a cluster of metal tubes about 3.5 metres long into which the fuel has been loaded. The fuel may be in the form of hundreds of uranium oxide pellets, each about the size of a pill, or it may be a solid rod or uranium which fits inside the tube. Each bundle can be safely handled on its own, since the fission process will not begin until the bundles have been assembled closely enough to form the critical mass of uranium needed for a chain reaction. The part of the reactor that houses the fuel bundles is known as the core. This is encased in a steel container with walls that can range from 23 to 30 centimetres thick and there is a further outer shielding of reinforced concrete several metres thick to ensure that there is no leakage of harmful radiation produced during fission.

Normally the stray neutrons given off in fission would travel at such speed that they would fly past other uranium atoms before they could bring about a further reaction. However, the stray neutrons can be slowed down if they are forced to collide with stationary atoms of a different type, contained in a

This refuelling machine at a British power station is 20 metres high and weighs 480 tonnes. The structure is suspended from a mobile gantry along which it moves on rails, and can be positioned over any one of 240 fuel pipes, here covered by square floor panels.

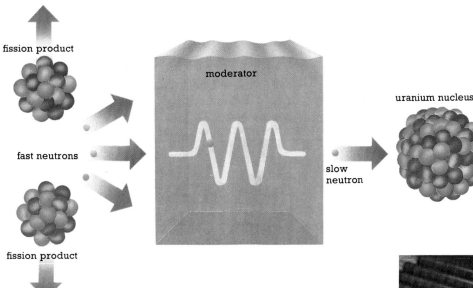

fission product

fast neutrons

fission product

moderator

slow neutron

uranium nucleus

This diagram shows how a neutron flying out of a split uranium nucleus ricochets through a moderator composed of graphite, water or heavy water. The braking effect of the moderator increases the chance of neutrons penetrating further nuclei (right) and triggering off a nuclear reaction.

substance known as a moderator. Suitable moderators are composed of atoms with only a small number of protons and neutrons in the nucleus, whose stronger inner bonds will not be broken by the stray neutrons. These neutrons will merely bounce off, losing speed in the process. Even so, these 'slow' neutrons still travel at about 7700 kilometres per hour.

The heat generated by fission can be removed from the centre of the reactor by using a liquid or gas known as a coolant. This passes over the fuel bundles and is then piped through a boiler where it turns into high-pressure steam for the generating process. The rate at which heat is produced by the reactor is controlled by rods placed between individual fuel bundles. These are made of materials which are able to absorb stray neutrons and so reduce the number of particles able to cause fission. Enough control rods will be installed in a reactor to completely stop the fission process if necessary. Usually a reactor is supplied with more than the minimum amount of fuel needed to produce a chain reaction, and heat production in the early stages is controlled by using the rods to absorb surplus neutrons. By gradually withdrawing the rods as fuel is burned a constant temperature is maintained.

Bundles of nuclear fuel can continue producing energy inside a reactor for several years before they need replacing. Although all the useful uranium may not have been used up, the new products formed during the fission process often include

Uranium oxide fuel pellets are carefully measured before being loaded into stainless steel tubes to form a fuel element cluster. Here a pellet is being checked for size on a micrometer gauge.

atoms which will compete with the uranium to absorb the neutrons which cause fission. This makes the fuel bundles progressively less efficient.

Because used fuel bundles contain substances that emit harmful radiation, they have to be carefully removed from reactors by remote controlled machinery. The bundles are transferred first to a cooling pond within the reactor building where they will be left underwater for several weeks. This gives their radioactivity and temperature time to die down. Other machinery may then be used to lift the bundles into a heavily shielded container for a journey by road, rail or sea to a plant where unused nuclear fuel can be separated from the harmful fission products.

Nuclear Reactors

The world's first nuclear reactor producing electricity for public use went into service at Calder Hall in Britain in 1956. It used graphite, which is a pure form of carbon, as its moderator and the heat was taken off by passing a coolant of carbon dioxide gas over the fuel elements and then pumping the heated gas over pipes containing water which was turned into steam. The uranium fuel was contained in tubes made of an alloy of magnesium which did not oxidize in the coolant, and the reactor was called the Magnox – **magn**esium alloy non**ox**idizable. Nine other Magnox power stations were built in Britain and the design was used for nuclear power stations built in Italy and Japan.

British nuclear engineers used experience gained in building the first Magnox reactors to come up with an even better design. The new reactor still used graphite as a moderator and extracted heat by carbon dioxide but it was also able to use fuel with a higher content of uranium 235 to give a heat output of 800 degrees Centigrade, compared with 400 degrees Centigrade temperature produced by the Magnox stations. More concentrated uranium fuel can also be kept longer inside the reactor which means that less frequent changes are needed. This improved design was called the advanced gas-cooled reactor, or AGR, and is still being built in Britain.

The United States began commercial nuclear power production a year after Britain. The first station was at Shippingport in Pennsylvania and started up in 1957. Like most reactors built in the United States since then, it uses ordinary water at high pressure as a moderator and coolant. The design is known as the pressurized water reactor or PWR and has been adopted by many other countries throughout the world as the basis for their nuclear power programme. The fuel bundles of the PWR are contained in a pressure vessel of thick steel, through

Advanced Gas-cooled Reactor

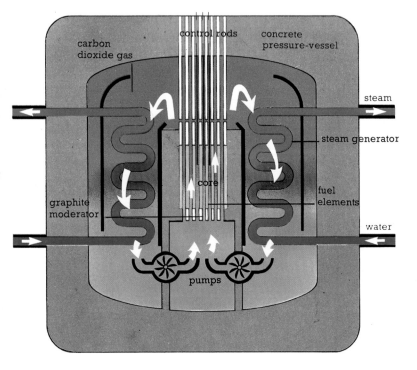

Pressurized Water Reactor

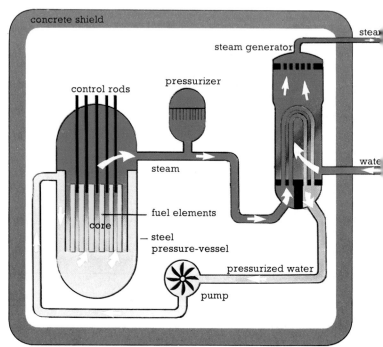

which water is pumped at high pressure. This slows down the neutrons to a speed at which fission will take place and draws off the resultant heat. Because the water is under high pressure, it does not boil at the normal temperature of 100 degrees Centigrade but emerges from the pressure vessel at more than 310 degrees Centigrade. It is then passed through pipes which transfer the heat to a fresh supply of water. This boils at about 150 times atmospheric pressure to form steam for electricity generation.

Another American reactor design, known as the boiling water reactor or BWR, uses some of the water from the moderator to produce steam which is piped straight to the generating plant. This gives a steam temperature of about 285 degrees Centigrade and has proved less popular than the PWR. One of the main advantages of both PWR and BWR designs is that the steel-lined pressure vessels containing the fission process, which are costly to manu-

facture, are much smaller than the size of vessels needed in British reactors using carbon dioxide gas coolant. A typical PWR needs a core only about 3.6 metres high and with a diameter of 4.5 metres, compared with the core 8.5 metres high and 9 metres in diameter needed for an AGR with equivalent output.

The Canadian nuclear power industry has developed another type of reactor in which the fuel bundles are placed in tubes through which the coolant is pumped. The design, known as CANDU, makes use of fuel with the small concentration of uranium 235 found in nature but is highly efficient because both the moderator and coolant are composed of heavy water. Heavy water is a combination of an atom of oxygen and two atoms of deuterium, which looks and tastes like ordinary water but is nearly 30 times more effective as a moderator. Besides being used in Canada, CANDU reactors have also been built in India and Argentina.

Boiling Water Reactor

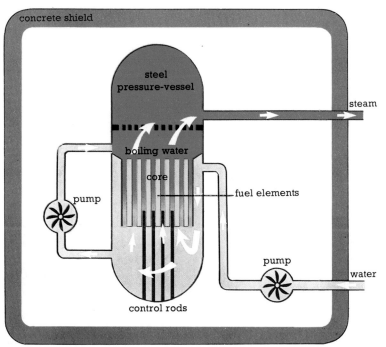

CANDU Thermal Reactor

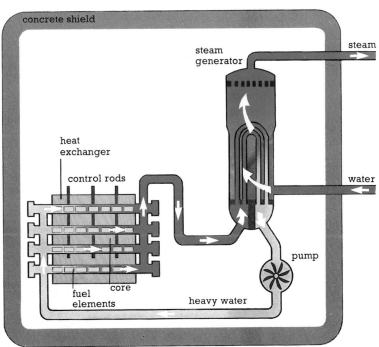

Nuclear Fuel

Uranium occurs naturally in many parts of the world. It is found in rock or sand deposits which can be mined on the surface or underground in the same way as coal or oil sands. The best deposits contain uranium oxide – a combination of uranium and oxygen – in proportions of about 25 parts to 10 000 parts of the ore, but in some deposits the concentration is as low as 25 parts in 100 000. Seawater contains about one part of uranium in 300 million, and a way may one day be found of making its extraction commercially worthwhile.

When uranium has been mined, it is treated with chemicals to dissolve and wash out the oxide. The solution can then be dried out into a powder which miners call 'yellow cake' because of its colour and floury appearance. At this stage only about seven in every thousand uranium atoms consist of uranium 235, the type most useful in nuclear reactions. The bulk of the uranium in the oxide is made up of uranium 238 which has a different balance of neutrons and protons in its nucleus, and so will not split and give off energy in the same way as uranium 235 when bombarded by relatively slow neutrons in a reactor.

Some reactors, like the original British Magnox design and the Canadian CANDU model, are designed to operate on natural uranium fuel, but most modern nuclear stations are designed for a higher output which is obtained from an 'enriched' fuel in which the proportion of uranium 235 has been artificially increased. It is possible to increase the proportion of uranium 235 to more than 95 parts in 100 but this is only done to give a very rich combination for use in the most powerful atomic weapons. Most nuclear reactors running on enriched uranium require a concentration of uranium 235 of less than four parts in 100 but even to achieve that level requires long and complicated processing.

Because enriched uranium can be used to manu-

At a uranium mine in Wyoming, heavy earth-moving equipment is dwarfed by the scale of the excavations.

Above: powdered uranium ore forms a fine, crumbly substance known as yellow cake, seen here adhering to a processing drum.

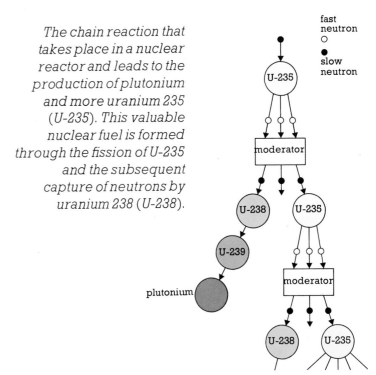

facture nuclear weapons, details of many of the techniques used are kept secret. But the main method of enrichment combines uranium with fluorine to produce a material called uranium hexafluoride, which is solid at room temperature but can be easily vapourized into a gas. If the gas is passed through a porous membrane containing tiny holes, the mixture that emerges on the other side will contain a slightly higher proportion of the lighter uranium 235 atoms. This needs to be repeated more than one thousand times to provide fuel with as low a concentration of uranium 235 as three parts in 100. The operation also requires large amounts of electricity – a little more than 5 per cent of the power that will eventually be generated by the fuel.

Even the richest nuclear fuel still contains a large proportion of uranium 238, but this is not entirely wasted in the reactor. Sometimes it captures one of the slow neutrons released by the fission process and turns into a substance known as uranium 239. This form of uranium does not undergo fission but instead begins to form another material called plutonium. Plutonium can also undergo a fission process if bombarded by stray neutrons, and so forms another source of nuclear fuel.

Plutonium and unused uranium are separated from other materials produced during the fission process by dissolving the spent fuel in acid and putting the solution through a series of processes in which separate materials are extracted. Recovered uranium can be mixed with fresh supplies to form new nuclear fuel while the plutonium is put aside as fuel for another type of nuclear reactor.

Much of the controversy surrounding nuclear power results from the problem of storing the ever increasing quantities of waste material that will be produced if more reactors are built. Although some scientists claim that highly radioactive materials can be buried deep in the Earth's crust without risk of contamination, others doubt whether it will be possible to prevent the radiation from escaping at some time in the future. Also, because the material stays dangerous for hundreds of years the problem of safe storage will be inherited by future generations, which will have to understand the danger of disturbing radioactive waste produced during the period when nuclear power was an important source of energy on Earth. One of the biggest challenges now facing nuclear scientists is the task of finding ways of dealing with large quantities of nuclear waste so as not to endanger future civilizations.

Fast Reactors

Plutonium produced during nuclear fission provides a way of extracting much more energy from uranium than is possible in a conventional reactor. If enough plutonium is assembled in fuel bundles within a reactor, a chain reaction will begin similar to that created by the fission of uranium 235 when additional neutrons are produced to split other atoms. The chain reaction in a reactor fuelled by plutonium not only produces enormous amounts of energy, but also releases plenty of fast-moving neutrons that will react with uranium 238, the most abundant form of uranium. This sets off a process known as breeding which eventually produces more plutonium. The transformation of uranium 238 into plutonium is much more rapid in this type of reactor than in a conventional design. Indeed, more plutonium can be taken out of the reactor than is put in with the fuel bundles.

The reactor used for this process has a special type of core that contains stainless steel fuel bundles loaded with a mixture of plutonium and uranium 238. During the fission process the uranium will be turned into fresh plutonium which can be used to provide further supplies of fuel. In this way, these reactors enable 60 to 80 times as much electricity to be generated from available supplies of natural uranium than would be possible using ordinary reactors.

The structure of reactors using plutonium differs from conventional reactors in an important way.

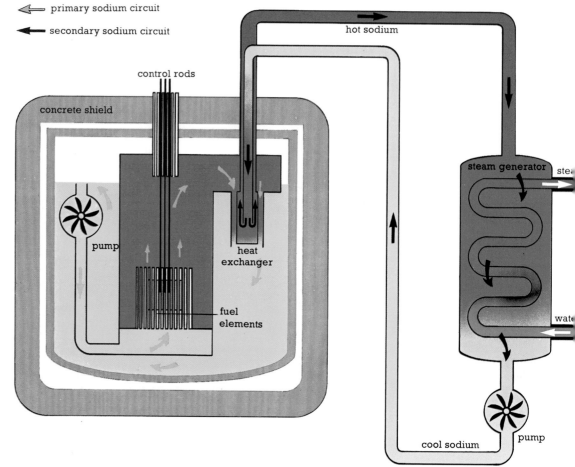

The core of a breeder reactor differs from conventional reactors in containing no moderator. Liquid sodium heated by the core passes through an intermediate heat exchanger where its heat passes to a secondary sodium circuit, which in turn transfers its heat to water in a steam generator.

primary sodium circuit

secondary sodium circuit

hot sodium

control rods

concrete shield

steam generator

ste

pump

heat exchanger

water

fuel elements

cool sodium

pump

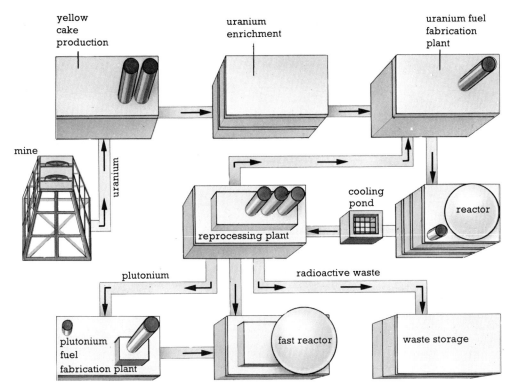

yellow cake production

uranium enrichment

uranium fuel fabrication plant

mine

uranium

reprocessing plant

cooling pond

reactor

plutonium

radioactive waste

plutonium fuel fabrication plant

fast reactor

waste storage

A specialist in the control room of a breeder reactor, watching the video console that helps to ensure that every aspect of reactor operation is constantly monitored.

This diagram shows some of the stages in a fuel cycle from the mining of uranium to the reprocessing and disposal of spent fuel.

They have no moderator to slow down the movement of neutrons produced during fission. This is why the reactor is often described as 'fast'. As a result a much more rapid chain reaction takes place, which allows the neutrons to escape into the uranium 238 around the plutonium and begin the breeding process. However, the temperatures generated in the core are much higher than in most other reactors, so a special coolant has to be used to extract the heat.

The substance normally used as a coolant is sodium, a metal which melts at a relatively low temperature of 99 degrees Centigrade and which in liquid form is ideally suited for use in fast breeder reactors. A pump fitted inside the reactor circulates the liquid sodium around the core and then through pipes where the heat is transferred to another stream of liquid sodium. This second stream of sodium leaves the container vessel and flows through more pipes where the heat will be transferred to water, producing steam. The rest of the process of generating electricity by a fast reactor is similar to that at other types of power station.

As in any nuclear power plant, the fuel bundles have to be taken out of a fast reactor after a time so that unwanted fission products can be removed. When this happens, the newly formed plutonium can be extracted and then used to provide a new source of reactor fuel.

Small power stations using fast reactors have been operating in Britain and France since the mid-1970s. The Soviet Union and East Germany are building similar stations and other countries are expected to follow as they amass stocks of plutonium and uranium from the reprocessing of fuel bundles removed from ordinary nuclear reactors. In the United States, however, the development of fast reactors has been slowed down by the argument that plutonium stocks might get into the hands of terrorists who could use them to manufacture nuclear weapons.

The development of fast reactors should lead to the setting up of a nuclear fuel cycle in which natural uranium will go into ordinary reactors to produce small quantities of plutonium which will be fed into fast reactors to produce more plutonium for recycling. In this way the maximum amount of energy will be recovered from the world's uranium resources to ensure a steady supply of electricity well into the twenty-first century.

Nuclear Security

Fission chain reactions not only produce huge amounts of nuclear energy. They also result in a rearrangement of protons and neutrons within the nuclei of uranium and plutonium atoms to produce new materials known as fission products. Many of these materials have too many particles within the nucleus to provide the normal balance between protons, neutrons, and electrons. As a result they begin to give off excess particles in a process known as radioactive decay, which can continue for thousands or even millions of years. The emission of particles is accompanied by radiation.

There is nothing strange or unnatural about radiation. Every day a stream of similar energy beats down on the Earth from the Sun, and although much of it is absorbed as it passes through the Earth's atmosphere there is still some that reaches us on Earth. Radiation is also produced by the radioactive decay of rocks containing naturally occurring uranium within the Earth's crust. Normally our bodies can absorb the small amounts of natural, potentially harmful radiation penetrating our skins and interfering with the atoms that make up the cells forming tissue. But exposure to such radiation can cause certain cells of the body to divide and grow in an uncontrolled manner, resulting in cancer, which is very often incurable.

The cores of nuclear reactors are encased in a steel armour surrounded by a layer of concrete some metres thick. This shielding is so effective that nuclear power stations give out less than one hundredth of the radiation we are receiving from natural sources, despite the huge concentrations of radiation they contain. Even so, there are fears that the increasing use of nuclear power stations could some day lead to an accident that would cause a massive release of radiation into the atmosphere comparable to the detonation of an atomic bomb. However, nuclear power stations are designed to release energy slowly and they cannot ever explode like an atomic bomb.

All nuclear power stations use remotely controlled machinery to ensure that workers are protected

The internationally agreed symbol to warn of the presence of dangerous levels of radiation.

A container that will be used to transport nuclear material is shown here being tested for its ability to remain intact after a drop onto concrete. The lead and steel box is surrounded by a wooden cladding that acts as a shock absorber.

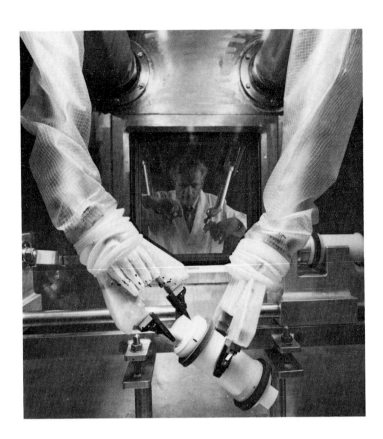

Shielded by a 'window' of zinc-bromide glass about one metre thick, a scientist uses remotely operated manipulators to open jars of radioactive fuel.

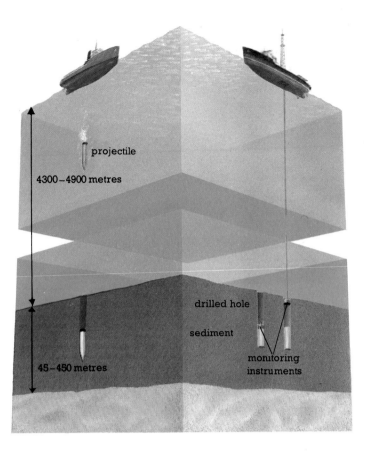

projectile

4300 – 4900 metres

drilled hole

sediment

45 – 450 metres

monitoring instruments

Shown here are two recently proposed ways of disposing of nuclear waste: on the left torpedo-like projectiles containing spent fuel are sunk in the ocean floor; on the right drums of waste are lowered into shafts by a winch. The burial process would continue as fresh ocean sediments gradually accumulate.

from highly dangerous substances, and waste materials may be transported to reprocessing plants in specially insulated containers. These containers, known as transport flasks, have walls of steel and lead and are so heavily constructed that a flask weighing 50 tonnes only carries about two tonnes of waste. To make sure that each flask is absolutely secure exacting tests are carried out involving exposure to temperatures of up to 800 degrees Centigrade and a 10-metre drop onto concrete.

At reprocessing plants, the waste materials are normally reduced into a highly concentrated volume of liquid and then put into stainless steel tanks with walls more than 2.5 centimetres thick which are in turn surrounded by thick concrete. The tanks are designed to be exceptionally strong, and are buried underground in concrete vaults as an added precaution. They also need to be cooled by piping cold water around them to pick up the heat generated by

the decay of the radioactive products.

Less than one part in 80 of uranium fuel ends up as waste material, but the amounts of material that will have to be stored safely for centuries will increase as more nuclear power stations are built. There have been suggestions that radioactive waste should be buried in uninhabited parts of the polar regions or even fired into space by rocket. But a more practical and less expensive means of disposal might be to turn the waste liquid into glass-like blocks which could be embedded in stainless-steel containers and buried deep in the Earth's crust or under the ocean floor. This would avoid the danger of water seeping through and picking up radioactivity that might spread to seawater and the soil.

Fusion Power

The nuclear energy we use today is produced by splitting atoms containing more than 200 particles in their central nucleus to release the forces that bind these particles together. However, there is another form of nuclear power that uses the energy released when much simpler atoms with just a few particles join together to form a different substance. This process, known as nuclear fusion, has been going on for billions of years inside the Sun and other stars, and has already been copied on Earth to produce hydrogen bombs which are perhaps the most deadly form of weapon. If the process could be harnessed for peaceful purposes it could provide an abundance of energy for centuries to come.

Nuclear fusion depends on supplies of deuterium

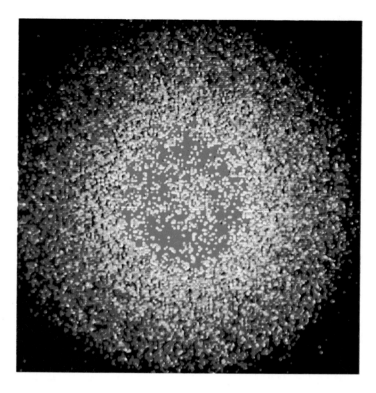

This unique photograph, taken in two trillionths of a second, shows a minute pellet containing deuterium smashing into fragments as it undergoes bombardment by laser beams in the core of a fusion chamber at the University of Rochester, New York.

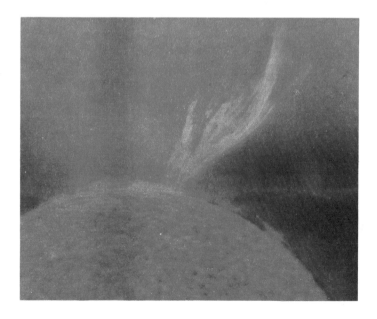

The glowing tongue of a solar prominence, in which temperatures reach 5500 degrees Centigrade, is a naturally occurring example of the plasma in which fusion reactions can take place.

and tritium, the special atoms of hydrogen which have neutrons as well as protons in their nuclei. Using electrolysis a cubic metre of water would yield enough natural deuterium to produce as much energy as 150 tonnes of oil if fusion power is successfully developed. Unlike nuclear fission, which produces radioactive waste, a nuclear fusion reaction produces materials that are relatively safe. These advantages have led many scientists to believe that fusion power could ultimately become the world's chief source of energy. But before this can happen tremendous obstacles will have to be overcome.

Man-made fusion reactions will only take place at incredibly high temperatures of at least 100 million degrees Centigrade which are six to seven times higher than those inside the Sun. At these temperature levels, deuterium and tritium atoms form plasma, a form of matter consisting of an equal number of highly charged positive and negative particles similar to

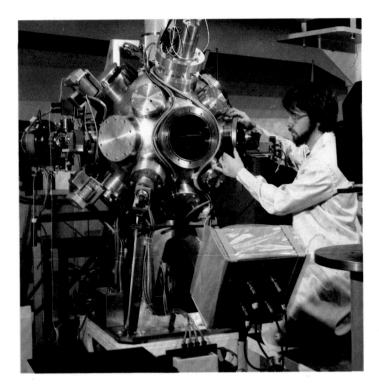

The target chamber in which the reaction on the left took place. Six laser beams can create temperatures of up to 100 million degrees Centigrade, and have produced one of the highest energy yields for input power ever achieved in laser fusion.

the material which creates the glow in the Sun. The atoms that make up the plasma are stripped of their electrons, allowing the positively charged nuclei to fuse together. For example, when a nucleus of deuterium in the plasma collides with a nucleus of tritium, they may unite to form a nucleus of helium. Because the helium contains fewer neutrons than the nuclei of deuterium and tritium, a free neutron, carrying an enormous amount of energy, is released.

This neutron can be captured by a metal called lithium, which as a result changes back into tritium and can be used in further fusion reactions.

Experiments have been carried out in the United States, the Soviet Union, Europe and Japan, as a first step towards building fully operational power stations using fusion reactors. One of the biggest problems is finding a way of containing the deuterium–tritium fuel since if the plasma touches the reactor walls it becomes too cool for fusion to occur.

One way in which this difficulty can be overcome is to use magnetic force to confine the plasma. Fortunately plasma is a good conductor of electricity and is influenced by magnetism. Russian scientists have developed a doughnut-shaped containment vessel surrounded by powerful magnets which may keep the material in place long enough for the reactions to occur. Reactors of this design are known as tokamaks and are being used as one basis for an on-going world research programme on nuclear fusion. In the United States and elsewhere fusion power research has also concentrated on creating fusion by aiming a series of high-powered laser beams at a pellet of the fuel positioned in the centre of the reactor chamber. The very hot beams create a temperature in which fusion will take place and because they are aimed at the fuel from all directions they will produce a force like a magnetic field to keep the plasma in place.

In trials that have been carried out so far, more energy has been needed to create magnetic fields and heat the plasma than has been given out by the fusion process. Because of this much more research is needed. Full-size fusion reactors will probably not be built anywhere until at least the first quarter of the twenty-first century and some scientists still doubt if the system will ever be made to work efficiently. But the prospect of converting water into energy is enough to encourage continued research.

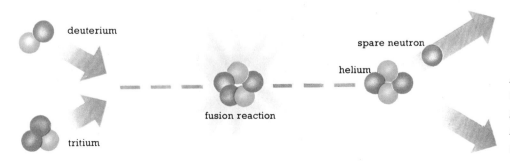

At about 100 million degrees Centigrade, deuterium and tritium nuclei fuse to form a nucleus of helium, releasing a spare neutron and an enormous amount of energy.

Storing Sunshine

Only about 50 to 70 per cent of the energy which the Sun beams towards the Earth actually reaches the ground. About a third is reflected back into space by the upper layers of the Earth's atmosphere, and more is absorbed as it passes through the rest of the atmosphere. Clouds also reduce the amount of solar radiation that reaches us on the surface.

Because the Earth's surface is curved, the amount of solar energy that manages to reach the surface of our planet varies from place to place. Rays of Sun falling on countries near the Equator have the least amount of atmosphere to pass through and therefore stay relatively strong, while rays reaching polar regions are considerably weakened by passing through more of the atmosphere. But even in a comparatively northern and cloudy region the amount of solar energy reaching the ground can be many times a nation's total energy requirement.

Unfortunately solar energy is only directly available when it is least needed – by day, when the atmosphere is relatively warm and the demands for other forms of heating are lowest. Similarly, in summer there is a smaller demand for artificial forms of heat than in winter, when there is much less solar radiation available. So if the Sun's heat is going to be useful to us we need to be able to convert it into a form of energy that can be stored.

One method of storing solar energy is by using the Sun to heat water that can then be kept in insulated tanks. The water is heated by passing it through hollow panels installed on the roof of a building. These panels, which work like household radiators in reverse, are usually made of materials – for example, black-coated steel plates – which absorb as much of the Sun's heat as possible and then transfer it to the water circulating inside. In many places there is enough solar radiation falling on the roof of a house even on a cloudy winter day to heat water sufficiently for washing and bathing. But there will probably need to be some extra source of heat, such as oil or gas, to provide hot water for radiators which can keep the house warm overnight. Even so, the use of solar energy to partially heat the water reduces the amount of fuel that would otherwise have been needed.

Solar panels generate relatively low temperatures, and heat produced in this way is described as 'low grade'. However, the Sun's rays can be concentrated into 'high-grade' heat, intense enough to convert water into high-pressure steam which can

This solar house at Malters, Switzerland uses energy from the sun to provide 70 per cent of its heating requirements. Maximum exploitation of this energy is ensured by building with wood, which is a better insulator than brick or concrete.

then be used to turn electric generators. The concentrated beams of sunlight are collected in a device known as a solar furnace which works rather like an enormous magnifying glass focusing all the sunlight falling on a large area of lenses or mirrors to produce a very high temperature at a single, small point. Experimental furnaces have been set up in various parts of the world to test ways of doing this on a large enough scale to provide steam for power stations. For example, complex arrays of mirrors can be powered by machinery to follow the path of the Sun across the sky, directing an intense beam of heat toward a boiler at the centre of the furnace from dawn to dusk. Besides providing electricity, this technique can create the high temperatures – over 3000 degrees Centigrade – necessary to manufacture certain types of pottery and tiles.

By concentrating sunlight collected by a battery of mirrors (below) temperatures of over 3000 degrees Centigrade are created in this solar furnace (above) at Odeillo in the French Pyrenees.

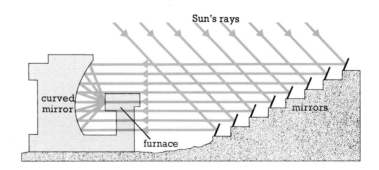

Power Stations in Space

▨ positive conductor	*This cross-section of a*
▨ copper sulphide	*photovoltaic cell shows the*
☐ cadmium sulphide	*sandwich of copper*
▨ negative conductor	*sulphide and cadmium*
▨ insulated backplate	*sulphide that releases an*
	electric current when
	exposed to light.

cover

Solar panels and furnaces simply collect and concentrate energy from the Sun. However, another type of collector, called a solar cell, uses sunlight to produce electricity directly, dispensing with the need for a conventional electric generator. The key to this process, known as photovoltaic conversion, is a material such as crystalline silicon or cadmium sulphide, which will give off a flow of electrons to an adjoining layer of metal when struck by sunlight.

The amount of current given off by a single cell is very small, and huge areas of valuable farmland around towns and cities would have to be covered with solar cells if the electricity produced in this way were to match the output from a more conventional power station. Deserts and other wastelands might be used instead but they are usually far away from the places where electricity is needed and costly power lines would have to be built. Solar cells are also very expensive to manufacture and cheaper methods will have to be found before the electricity they generate can compare in price with that produced from coal and nuclear power, especially since conventional power stations continue operating at night when there is no sunlight to work the cells.

Photovoltaic conversion is most useful in situations where conventional ways of producing electricity are impracticable. For example, solar cells can be successfully used to provide electric power for satellites and space missions. In 1973 and 1974 United States astronauts spent many months orbiting

Huge solar collectors, based on the photovoltaic cell and up to 115 square kilometres in area, could one day convert the Sun's energy into electricity and transmit it back to Earth as radio microwaves.

A laboratory in California has developed this solar-powered buggy in which an array of cells affixed to the roof charges batteries supplying a ten horse-power electric motor. The vehicle has a maximum range of 65 kilometres and could serve as a prototype for solar-powered transport on the Moon.

and NASA. The most familiar is the construction of large arrays of cells. Such collectors would be too big to be carried ready-made from factories on Earth, so they would have to be constructed from components and materials carried on board space-craft. However, this would not be as difficult as it might seem because the weightlessness of objects in orbit around the Earth would make it relatively easy to manoeuvre huge pieces of the structures into place. The technology now exists to enable these orbiting power stations to be built and by the mid-1980s the American space shuttle will be available to carry construction materials out into space.

Solar power satellites would be placed in orbits which kept them above the same point on the Earth's surface all the time. The solar power would be converted into a microwave beam, the same high frequency radio signal that is sometimes used in ovens to cook food. This could be picked up on Earth by dish-shaped antennae covering an area of about 100 square kilometres, and then be converted back into electricity for the public supply system. The intense heat of microwaves would scorch the Earth's surface if a beam from space were to go off target, but scientists believe that they can ensure that the transmissions will automatically shut down if the beam is thrown off course.

the Earth in the satellite Skylab, whose array of solar collectors produced much of the electricity that was needed. Cells work much more effectively in space than on Earth because of the almost continuous stream of sunlight available, and this has led scientists to suggest ways of using satellites to generate electricity in space and beam it back to Earth.

To produce worthwhile amounts of solar energy in space very large satellite collectors would be needed, and a variety of systems have been studied in the United States by the Department of Energy

The United States space shuttle Columbia *prepares to land at Edwards Air Force Base, California, after its first space voyage in April 1981. Craft of this type may one day ferry materials for constructing arrays of solar panels in space.*

Heat Pumps

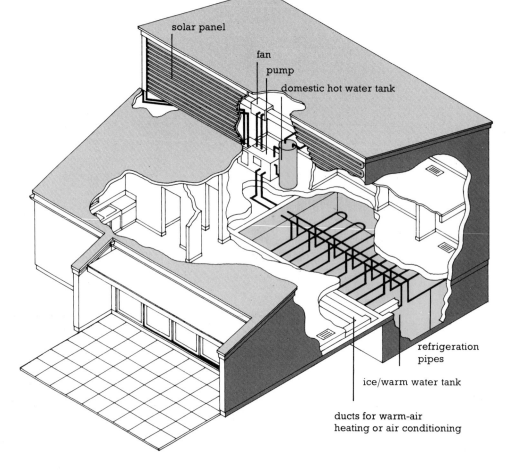

solar panel

fan
pump
domestic hot water tank

refrigeration pipes

ice/warm water tank

ducts for warm-air heating or air conditioning

The annual cycle energy system (ACES) house in Knoxville, Tennessee, contains a large water tank in its foundations. Ice formed from cold air extracted from rooms in winter provides air conditioning during the summer months. A solar panel helps to boost hot water supplies.

We usually think of energy as supplying heat, but it can also cool things down. A household refrigerator is an example of energy – usually electricity – being used in this way. The power supply is used to pump a liquid with a low boiling point, such as ammonia, through pipes surrounding the food cabinet. Such a liquid will easily absorb heat, in the same way that a pan of water will pick up heat from a burner, and will quickly turn into a vapour. As it leaves the food chamber the gas is compressed back into a liquid while passing through a coil of piping known as a condenser, and heat is given off to the atmosphere before the liquid is piped through the cabinet again to extract more heat.

A similar method is employed to cool the inside of a building by using a fan to drive warm air past a refrigeration unit which absorbs its heat. This process can be reversed to extract heat from the air, or even the ground outside a building and transfer it inside. A system able to move heat from one place to another in this way is known as a heat pump.

Pumping systems have been designed for build-

ings, thus enabling the flow of heat to be reversed to keep rooms warm in winter and cool in summer. The only energy needed is electricity to drive the compressor, which converts the gas into a heat absorbing liquid and circulates it through the system. Even in winter sufficient heat can be extracted from rivers, lakes, or the ground to provide a cheap supplement to conventional heating methods.

Although heat pumps are unlikely to provide all the warmth needed for domestic heating in cool climates, by partially heating air or water before it is piped through heating systems they can reduce the amount of heat that needs to be produced by oil, gas, or electricity. Heat pumps can also be used to recover heat lost through chimneys, flues, and drains, and if cheap enough models can be produced they may become a familiar feature of home heating.

In the future it may even be possible to build houses with heat pumps that could be used in winter to extract warmth from a pond of water built into the foundations. This would be just like the refrigeration process used to produce skating rinks, where pipes

This experimental thermal-energy conversion system works by using warm water from the ocean surface to evaporate a volatile liquid, such as ammonia, which as pressurized vapour turns a turbine. Cold water from the ocean depths is used to condense the vapour, which is then recycled through the system.

The domestic refrigerator acts like a simple heat pump. A volatile liquid, such as ammonia or sulphur dioxide, is pumped through tubes lining the freezer compartment where it absorbs heat and cools the interior. The gas is then compressed and passes through condensing coils with fins which give out heat, where the gas condenses into a liquid and starts the cooling cycle again.

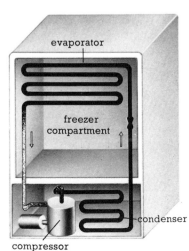

evaporator

freezer compartment

condenser

compressor

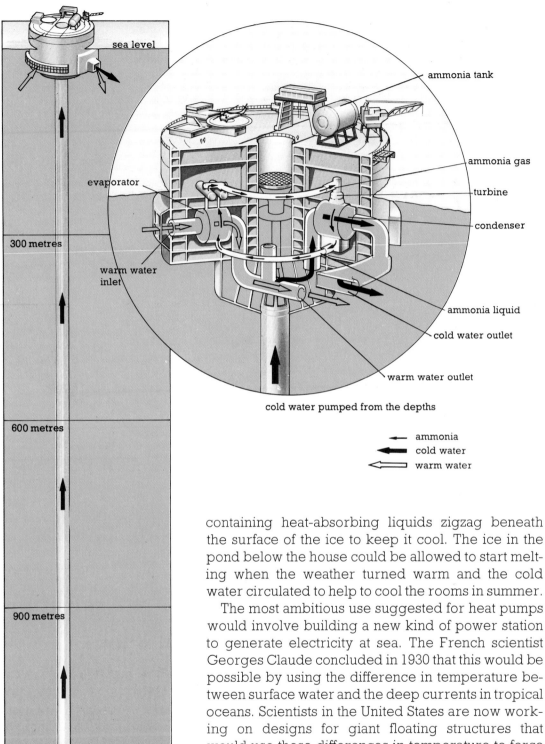

sea level

evaporator

300 metres

warm water inlet

600 metres

900 metres

1200 metres

ammonia tank

ammonia gas

turbine

condenser

ammonia liquid

cold water outlet

warm water outlet

cold water pumped from the depths

→ ammonia
→ cold water
⇦ warm water

containing heat-absorbing liquids zigzag beneath the surface of the ice to keep it cool. The ice in the pond below the house could be allowed to start melting when the weather turned warm and the cold water circulated to help to cool the rooms in summer.

The most ambitious use suggested for heat pumps would involve building a new kind of power station to generate electricity at sea. The French scientist Georges Claude concluded in 1930 that this would be possible by using the difference in temperature between surface water and the deep currents in tropical oceans. Scientists in the United States are now working on designs for giant floating structures that would use these differences in temperature to force liquids, like those used in refrigerators, to vapourize when warmed by surface water. The vapour would expand and spin turbines to generate electricity. Cold water brought up from the depths would be used to condense the vapour into a fluid again.

Underground Energy

The Earth has its own internal source of heat – a fact particularly well known to miners and oil explorers who find that temperatures rise as they search deep down in the Earth's crust for fuel supplies. This heat sometimes results in hot springs bubbling to the surface.

Hot springs occur when warmth from the Earth's interior penetrates through rock and heats water in underground deposits to above boiling point. This often causes a build-up of pressure that sends the hot water shooting out of the ground. Hot springs of this type are called geysers, and are found in earthquake and volcano areas of countries such as Italy, New Zealand, Iceland and the United States. The heat they contain is known as geothermal energy.

Even in areas not subject to earthquakes or volcanic eruptions, hot water or steam can sometimes be found trapped in porous underground rocks similar to those in which oil and natural gas occur. These supplies of hot water or steam will be kept constantly at a boil by heat within the earth's interior. However, because it is sealed deep underground at high pressure the water does not vapourize and will easily flow into a well drilled from the surface. As the water rushes to the top of the well the easing of pressure allows it to turn into steam, and this can then be used to drive electricity generating turbines, as does the steam produced by burning coal or oil. The world's largest geothermal power complex, at The Geysers, California, began operation in 1960 and by 1983 will have an annual energy output equal to that from almost 13 million barrels of oil – sufficient to serve a city of one million people with electricity.

It is usually not possible to pipe underground water directly into home heating systems because the chemicals and sediments it contains would clog up the narrow pipes. Instead the water is passed through a heat exchanger where it flows over pipes containing purified water from the public supply system. These absorb the heat and then pass the water on to household tanks and radiators.

Water from deep below ground is often difficult to get rid of once its heat has been taken off, since the high concentration of chemicals and sediments would cause serious pollution if discharged into rivers or the sea. The problem can be overcome by sending it back through a second well to the rock formation from which it was originally taken. Even so, care has to be taken to make sure that the outlet from the disposal well is not too near the extraction well. Otherwise the cooler water returning from the surface

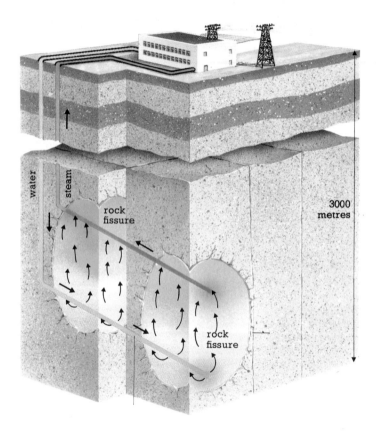

This diagram shows how steam can be generated from hot rocks deep in the Earth's crust. Cold water is pumped into a series of intersecting disc-shaped fissures created by hydraulic pressure. As it absorbs heat from the rock the water becomes less dense and rises back toward the surface at high pressure. If the pressure is then reduced the hot water turns to steam and can be used to turn the turbines of a power station.

might reduce the temperature of the fresh hot water.

Many parts of the world contain huge areas of rock which can become very hot beneath the surface but are too solid for water to penetrate. In the future it may be possible to tap this dry source of heat. By drilling wells and injecting high-pressure water engineers could fracture the rock, creating a series of disc-shaped fissures. Water at normal pressure could then be pumped down to pick up heat as it flowed through the fissures.

Steel pipes carry pressurized steam from deep underground to the generator of a geothermal electric power station in New Zealand.

River Power and Hydroelectricity

Anyone watching a river in flood realizes the enormous energy in running water. It is one of the oldest forms of power known to us. Farmers used to take their grain to mills driven by water power, and much bigger wheels were built to power factories in the early stages of the Industrial Revolution. Even today water power provides about five per cent of all energy used throughout the world.

Most of the mechanical energy now created from water is converted into electricity. This is generated by piping water through turbines which spin under pressure just as other kinds of turbines are turned by steam pressure in power stations burning coal or oil. Electricity produced by water pressure is known as hydroelectric power. Sometimes turbines can be placed directly in the currents of fast flowing rivers but more usually water is collected behind dams and fed through pipelines running down to a power station situated below the level of the dam wall. It is really the effect of the Earth's gravity that provides the power, and the farther the water has to fall from the dam head to the turbine the greater the power output that can be achieved. A very large hydroelectric station has been built in the Soviet Union to give an output nearly equal to that of eight typical nuclear reactors. In the United States, the Grand Coulee Dam provides power about equal to the output of three nuclear reactors.

Mountainous areas provide the best sites for hydroelectric plants, but these tend to be a long way from the main centres of population where power is needed. This means that electricity has to be sent long distances by overhead power lines, creating an eyesore in landscapes of great beauty. The hydroelectric dams can also spoil the appearance of upland areas. But with modern planning and construction techniques the problem can be lessened by installing some of the more unsightly facilities underground.

Sometimes there are too few rivers and streams in

The Glen Canyon Dam in Arizona is one of the key features of a scheme that has dammed the Colorado River to provide both hydroelectric power and irrigation. Its wall, 216.49 metres high and 475.48 metres wide, holds back a lake 300 kilometres long. Water passes out of the lake in pipes driven through the base of the wall and turns turbines in the power plant visible at the bottom of the picture. Great dams like this provide vast amounts of cheap electricity. But the resultant upstream flooding and silting has led some countries to slow down construction.

Water from a fast-flowing stream, guided along an artificial channel, gushes onto the blades of a wooden water wheel to provide energy for a mill at Lille Mølle, Denmark. Wheels of this type, which have been in use for centuries, have a total output of about one horsepower.

lakes feeding the dams, and this can vary a great deal from season to season. Fortunately, many countries using this form of electricity generation receive large amounts of rain in winter when demand for power tends to be greatest. Overall, hydroelectric schemes are able to meet about a quarter of the world's total annual demand for electricity.

This Pelton-type hydroelectric turbine, shown shortly after construction at a factory in Oslo, Norway, has a potential output of 330 000 horsepower, making it one of the most powerful single turbines ever to be built.

a hilly area to provide hydroelectric power in the conventional way. Such areas can, however, be used for so-called pumped storage complexes which consist of a dam built on a hillside and linked by pipeline to turbines in a power station some distance below. During the night the dam is filled with water pumped up from a lower level, possibly from a lake. This uses up surplus electricity generated at power stations operated by coal, oil or nuclear power, which cannot easily be switched off as demand for electricity falls when most of the population goes to bed. The water in the dam can then be released during the day to reinforce other supplies when there is a sudden increase in electricity demand.

One of the greatest advantages of hydroelectric power is that it can be switched on very rapidly, as long as there is a supply of water in the dam. In other types of power stations, steam turbines take time to warm up and reach peak output which makes them unsuitable for coping with sudden rises in power demand. However, the amount of hydroelectric power available depends on the level of rivers and

Waves, Tides and Turbines

Just as the flow of water in rivers provides energy that can be converted into electricity, the movement of water in the oceans represents a vast source of untapped energy. Tides, with their regular rhythm, are more reliable than the rainfall needed to replenish hydroelectric dams. If an estuary is bridged by an artificial barrier the incoming tide can be captured, creating a reservoir of water that can be used to generate electricity as the tide recedes.

Electricity is produced by an improved version of this method in a dam across the Rance estuary in north-west France. As the tide rises seawater generates electricity by passing through 24 turbines. When the high-tide level is reached the turbines stop. The receding tide then rushes back through the turbines with sufficient force to generate more hydroelectricity. The scheme also uses electricity generated during off-peak periods at other power stations to turn the turbines into pumps to boost the inflow or outflow of water, thus creating even more power.

Unfortunately, there are only two cycles of several hours each day when the tide is rising and falling and the turbines can produce sufficient power, so tidal dams of the Rance type cannot be relied upon to give a steady flow of current. There is also the disadvantage that the tides do not always coincide with the times when electricity demand reaches a peak and a supply of hydro power could back up the output from other stations. In spite of these problems tidal power has enough potential to be planned in other parts of the world where suitable estuaries occur.

There is also increasing interest in extracting energy from waves, which provide a more continuous source of power than tides. This has the added advantage that the strongest waves tend to occur during winter periods when the demand for electricity is also at its peak. Japanese engineers have devised a method for generating wave power using

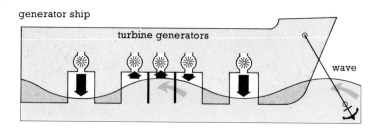

This diagram shows an experimental floating power station constructed in Japan. Turbines are sited above vents in the underside of the generator ship and are driven by the alternate suction and compression of air resulting from the rhythmic rise and fall of waves.

a structure similar to the hull of a ship. As the waves pass underneath the vessel, air is forced through turbines installed above vents in the lower part of the hull. If the vessel were anchored in a stormy area of sea, large amounts of energy could be captured in this way. Research is continuing in order to find the most suitable design of air turbine.

In the future wave power could also be harnessed by using a string of rafts or hollow concrete structures anchored along the line of the waves. As they are hit by successive waves, they will move backwards and forwards. This motion will be converted into a flow of hydraulic fluid which drives electric turbines. If this idea is developed on a large scale, electricity will be sent ashore by cables laid along the seabed.

If wave power is to become a reality, engineers will have to perfect the design of devices to ensure that they will be sturdy enough to survive exceptionally rough weather, while still being able to tap energy from the waves in normal conditions. Metal parts will need protection against the corrosive effects of the salt in seawater. However, problems could be created for shipping by long lines of low-lying rafts which might not be seen in rough or misty conditions.

The 760 metres long dam (right) across the Rance estuary in north-west France, where the tidal variation is often over 12 metres, contains the world's first industrial tidal power project. Two-way turbine generators (below left) are installed within a hollow section of the dam wall and are turned by both in-coming and out-going tides. The water flow can be halted by emergency sluice gates, while additional slots permit the installation of small dams for turbine maintenance.

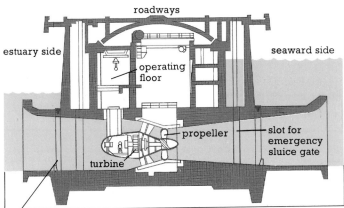

roadways

estuary side

seaward side

operating floor

propeller

slot for emergency sluice gate

turbine

slot for maintenance dam

A proposed wave power system developed by British scientists. As the outer frame is rocked backwards and forwards by wave motion it moves against the fixed inner core. At each cycle water pumped through the core (pale blue arrows) is sucked through one-way valves into cavities between the core and the frame and then compressed as the frame moves in the opposite direction. This results in a flow of high-pressure water (dark blue arrows) through a further set of valves and into lateral ducts leading to turbines further along the section.

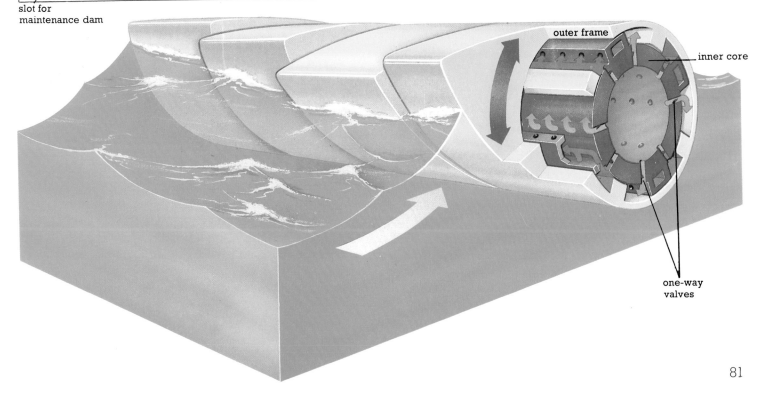

outer frame

inner core

one-way valves

81

Plant Power

Although most of the sunlight shining each day is lost back into space after striking the surface of our planet, about 0.1 per cent is trapped by the Earth's abundant plant life. This takes place through photosynthesis, in which plants absorb energy from the Sun and, with the aid of carbon dioxide and water, produce carbohydrates. This energy store has been used by people for centuries in the form of coal but we do not have to wait millions of years to enjoy all the benefits of energy trapped in vegetation. There are ways in which we can exploit the power in plants that are still growing.

The most direct way of getting energy from plants, apart from those we can eat to build up stores of energy in our bodies, is to burn vegetable matter as a fuel. Wood is one of the best sources of such heat and was once the only fuel known to us. It is still the chief form of fuel in many parts of the world, but traditional ways of burning vegetable matter in open fires are very inefficient because of the amount of useful heat that is lost to the surrounding atmosphere. New techniques are now being developed which promise to give far better rates of energy extraction.

One way of getting more energy from plants is to heat them under pressure in a boiler. If air and steam are added a rich mixture of gases, including hydrogen and carbon monoxide, will result and this provides a fuel that can be burned for heating and cooking. This, of course, requires supplies of energy to heat the plant material and to cultivate, harvest and transport the crops. The process must therefore be carefully controlled to make sure that more energy is obtained than is used to produce it.

Plant life that can provide useful forms of energy is known as biomass. Some plants are more suited to this than others. For example, sugar cane provides a useful fuel if its juice is fermented to produce alcohol, in much the same way as grape juice is allowed to mature into wine. Sugar cane fermentation has been developed on a large scale in Brazil where the climate is ideal for growing the crop. The fermentation is carried out in more than 300 distilleries which in 1980 had a combined output of more than 3.5

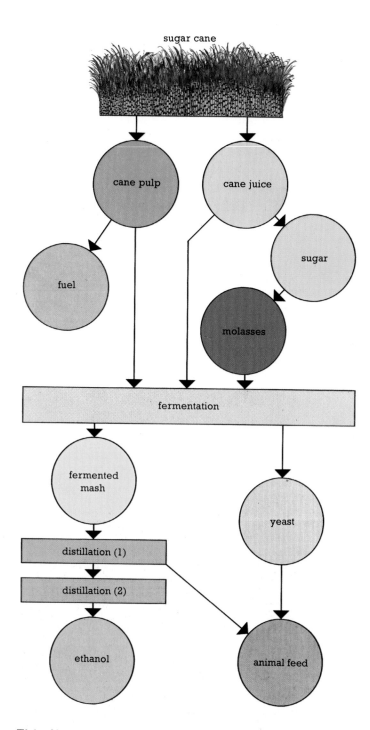

This diagram shows the elaborate process through which sugar cane is put, producing many by-products before energy, in the form of ethanol, is finally extracted.

billion litres.

At present a large proportion of the alcohol fuel produced in Brazil is added to petrol giving a high energy fuel known as gasohol. This can be used in conventional car engines, the most acceptable blend being one with an alcohol content of 20 per cent. However, increasing numbers of cars are now being modified to run on alcohol alone. Scientists in Brazil are also investigating ways of fermenting other crops – for example, cassava and sorghum – to provide alcohol while in other countries, particularly France, grain is being fermented to provide a petrol additive.

One of the difficulties about growing crops to produce energy is the amount of land that is needed to get a worthwhile yield of fuel. It has been estimated

Studies of water hyacinths, seen here growing in a tank at a NASA research laboratory in Mississippi, have proved their hidden potential as a source of methane. Despite their water cleansing properties, these plants are themselves a form of pollution, choking shallow tropical waters with dense mats of weed, and their harvest as fuel would remove a major hazard to fishing boats and coastal shipping.

Sugar cane growing in Brazil is being turned into a valuable alcohol called ethanol, used as a supplement to petrol and as a fuel in itself to power cars. Even so, Brazil must still import about 75 per cent of the petrol it consumes.

that even if an area of land the size of the state of Arizona were planted with special energy-producing crops, it would only produce enough fuel to replace one tenth of the supply the United States gets from conventional fossil fuels like coal and oil. There is also the drawback that energy crops take up land that could otherwise be used for food crops or recreation.

These problems may one day be solved by harvesting plants that grow under water. Certain forms of seaweed can be fermented in the absence of oxygen to produce a mixture of methane and carbon dioxide gas, and a similar process can be used to produce gas from the water hyacinth, which thrives in polluted rivers and lakes. Water hyacinths have exciting potential as an energy source, besides providing fuel they help to clean water by absorbing many of the pollutants.

Hydrogen Power

Most of the fuels we use contain a proportion of hydrogen atoms, the simplest of all elements, usually combined with carbon atoms. When the fuel is burned, the hydrogen breaks away and joins up with oxygen from the atmosphere to form water, releasing energy in the process. However, hydrogen forms a valuable fuel when not combined with carbon. It also has the unique advantage that after it has been burned to release its energy, all that remains is pure water. There are none of the polluting by-products that result from burning naturally occurring fuels, which usually contain impurities along with the hydrogen and carbon.

A jeep of the United States Postal Service has been specially adapted to run on hydrogen. The 1590-kilogramme vehicle has a range of about 113 kilometres at 88 kilometres per hour and has a maximum speed of 113 kilometres per hour.

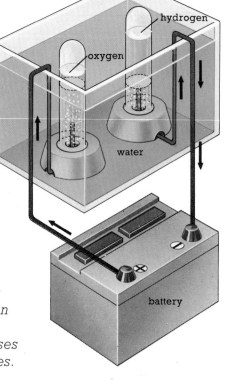

This diagram shows the process known as electrolysis: an electric current passes through water and breaks it down into hydrogen and oxygen atoms, which collect as gases at the top of the tubes.

Chips of iron-titanium hydride stored in a 225-kilogramme tank in the back of the jeep absorb hydrogen when cool and release it when heated by waste heat from the engine.

The reason we do not use more pure hydrogen as a fuel is that it is almost impossible to find it in its pure, gaseous state. Although hydrogen is very plentiful it is normally combined with other elements to produce materials such as coal, oil, or water, and very expensive processes are needed to separate it.

The hydrogen and oxygen atoms in water can be forced apart by a process known as electrolysis, which involves passing an electric current through the water and collecting the separated atoms at the points where the current enters and leaves. Hydrogen can also be separated from water by heating it at temperatures above 2700 degrees Centigrade. This technique, like electrolysis, requires an initial input of energy equal to that obtained from the pure hydrogen and illustrates the fact that hydrogen is a convenient way of storing energy rather than a means of producing it.

The development of nuclear reactors operating at very high temperatures could in the future provide a way of producing large quantities of hydrogen from water by the heating process. It could then be transported to homes and industry through networks of pipelines just as natural gas or coal gas are at present. The hydrogen could also be stored for transport purposes by freezing it into a liquid fuel, a process that greatly reduces the space that the fuel takes up. However, special care has to be taken with the materials chosen for storage because hydrogen atoms are so minute that they can pass through the space between the atoms of some metals, causing a dangerous leakage of this highly flammable element.

Ways are now being developed to store hydrogen in solid form in the fuel tanks of cars. In the United States experimental vehicles have been built with fuel tanks that can be partly filled with crushed particles of a metal able to absorb hydrogen gas. The metallic chips can store more hydrogen than the same amount of empty space, and act like a chemical sponge which soaks up hydrogen when it is pumped into the tank and then releases it when heated. Once the engine has been started – by a conventional electric motor – the chips can be activated by heat taken from the engine system, and they can be used time and again to absorb and release fuel.

Hydrogen can also be used to produce electricity by mixing it with oxygen in an enclosed chamber known as a fuel cell. This works rather like a con-

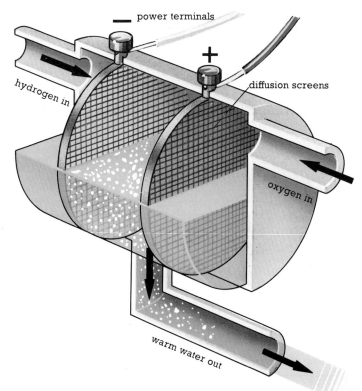

In this fuel cell, oxygen and hydrogen are fed through diffusion screens to a chamber where they combine to form water. The electrical energy released by the union of hydrogen and oxygen ions is collected at terminals attached to the top of the screens.

ventional battery, but contains only a substance known as an electrolyte – for example potassium hydroxide. Hydrogen and oxygen are fed into the cell and the electrolyte causes the two elements to combine, generating electricity in the process. The only by-products of this chemical reaction are heat and a supply of warm water, and although the energy obtained is always less than that needed to produce the hydrogen gas by electrolysis the fuel cell provides a very clean, convenient package of power.

Fuel cells are still at an experimental stage but in the future they may provide a better way of storing electrical energy than do conventional batteries, which take up a lot of space. Power generated overnight when there is little demand for electricity could be turned by electrolysis into hydrogen and oxygen and stored in liquid form. When demand increases the two elements could be passed through a fuel cell to provide an additional source of power.

The Return to Wind Power

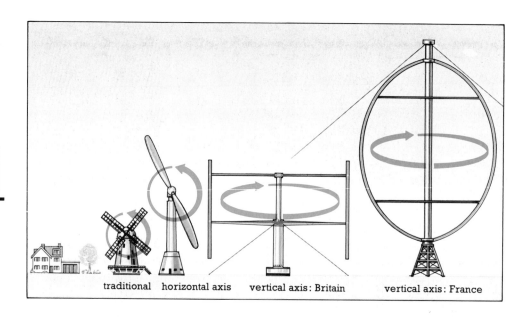

traditional   horizontal axis   vertical axis: Britain   vertical axis: France

The daily movement of air across the Earth's surface creates infinitely more energy than we normally need. For centuries this energy source has been used to grind corn and pump water, particularly in such places as The Netherlands, where there are no mountains with fast-flowing rivers to provide water power. Traditional windmills can still be seen in many parts of the world but now much bigger versions are being built to generate electricity.

Knowledge of aerodynamics gained from building aircraft has helped engineers to develop the most efficient design of blades for modern wind machines. The latest windmills have blades with a total length of up to 90 metres that look like giant aircraft propellers. These are installed on towers with a revolving top section which can be turned by a motor so that the blade can be set at an angle that will pick up the maximum amount of energy from the wind. The power is then transmitted through a shaft to a turbine to generate electricity.

In spite of the size of some experimental wind machines they produce very little electricity compared with power stations running on fossil or nuclear fuel, or even hydroelectric schemes. It would take several hundred windmills to replace the energy generated in a typical modern power station, and this would require an array of tall towers sited on a hillside where winds are stronger than at ground level. Many people would regard such a cluster of windmills as an eyesore, and the towers might also

Modern machines for harnessing wind power, left, dwarf the traditional windmill and tower above a two-storey house. A horizontal axis model is under construction in the Orkney Islands, Scotland, and will have a total length of blade greater than the wingspan of a Jumbo Jet. Of the vertical axis models the projected British design has the advantage of a higher tip speed for a given wind velocity.

interfere with radio and television signals.

Some scientists have suggested overcoming this difficulty by building windmills at sea, where there is the additional advantage of stronger, more constant winds. Offshore oil exploration has already developed ways of constructing towers that would stand on the seabed and support the blades above the waves in areas of shallow water. As long as the machines were far enough offshore, their visual impact would be minimal, although they might be a hazard to shipping.

Electricity generation by wind is only likely to be carried out on a large scale if engineers can build machines that are inexpensive enough to make this 'free' source of energy comparable in price with power generated by coal, oil, or nuclear stations. Otherwise, only small-sized wind generators are likely to be used, providing electricity at farms and ranches too remote to be linked to the public supply.

A three-metre windmill such as the one below, familiar to generations of farmers, can provide about one horsepower from a 30-kilometre-per-hour wind.

A vertical axis wind turbine built to the design of the French inventor Darrieus, right, shortly after completion in the Magdalen Islands, one of the windiest spots in Canada. The whisk-shaped blades of this giant turn around a shaft 36.57 metres high and form a loop 24.38 metres wide.

87

Energy in Demand

No form of energy comes to us free of charge. A price has to be paid for the fuels that warm our homes and give us light and power. We have to buy the petrol that runs our cars and when we travel by rail or air a large part of the fare goes to cover the cost of fuel. Even if we were to generate electricity by our own private windmill, we would not get the benefits of that natural source of energy for nothing. We would first have to pay for buying and installing the equipment and we would later face regular maintenance bills to keep it in good working order.

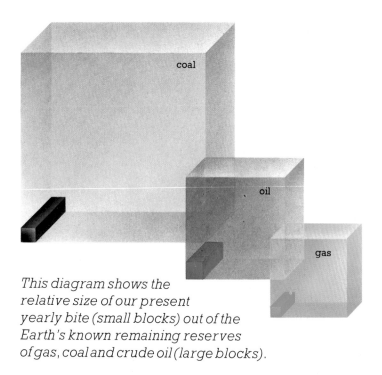

This diagram shows the relative size of our present yearly bite (small blocks) out of the Earth's known remaining reserves of gas, coal and crude oil (large blocks).

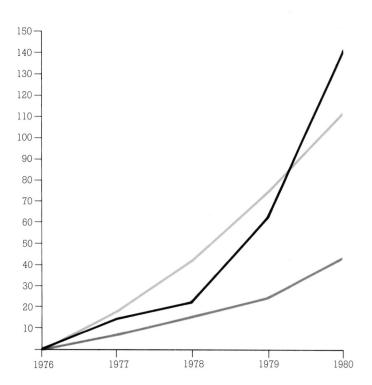

Oil price rises, shown by the rocketing cost of heating oil to domestic consumers (black), are reflected in rising costs of electricity (red) and gas (blue). The numbered scale serves as an index of price rises.

Many things affect the price we pay for energy, but the biggest single factor is its availability. Like any other commodity, energy becomes more expensive if there is any real or threatened shortage of supply. At such times, those who have some form of energy to sell want to get as good a price as possible for something they cannot replace once it is sold, while those who are rich enough will pay as much as they can afford if they are eager enough to buy it.

Most of the energy upon which we depend is exhaustible. This means that there are only limited supplies and the amount available for the future is reduced every time some of these are used. Fuels like coal, natural gas, and oil fall into this category. In contrast, wind power and solar energy are classed as renewable because they are replenished each day by the Earth's natural processes.

The modern world's main source of energy is an exhaustible one – oil. In recent years we have burned up more than 3 billion tonnes annually to provide about 45 per cent of our total energy, and if we were to go on using it at that rate we would exhaust the world's known oilfields in little more than 25 years.

Oil was once **very** cheap and as more of it became available, demand began to rise quite dramatically. Between 1950 and 1960 the amount used doubled and

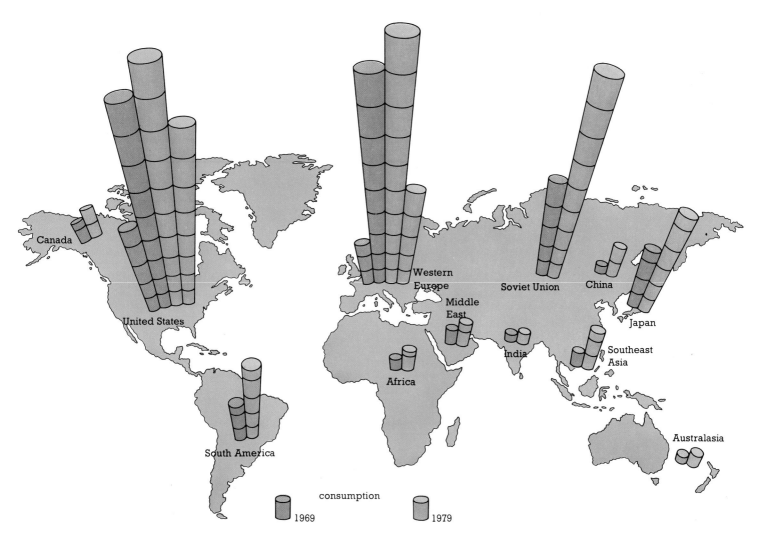

Canada

United States

South America

Western
Europe

Middle
East

Africa

Soviet Union

China

India

Japan

Southeast
Asia

Australasia

consumption

1969

1979

Each barrel on this diagram represents one million barrels of crude oil consumed daily. Despite steadily diminishing supplies and rocketing prices, world oil consumption rose by 50 per cent between 1969 and 1979. The trend was most marked in the Soviet Union, the Third World and China, with the last showing a staggering 350 per cent increase.

it more than doubled during the ten years up to 1970. This enormous increase in oil consumption meant that supplies were in danger of running out at a frighteningly rapid rate. In 1950, for example, the amount of oil left in known oilfields was equal to nearly 100 years' supply, yet by 1975 so much was being used that, given the same rate of consumption, only 40 years' supply remained. Of course, important new oilfields are still being discovered, especially in remote ocean areas, but it is doubtful if enough will ever be found to meet the present rate of demand.

The Earth's dwindling oil supplies are of great concern to the group of 13 nations which produce most of the oil used by the industrialized world. In 1960 these oil-producing countries decided to form a group called the Organization of Petroleum Exporting Countries, or OPEC, so that they could fix a common price for the oil they were supplying to the rest of the world. Many of the OPEC states are in the Middle East, which contains over half of the world's oil reserves and although the oil is quite cheap to produce it is sold at a price that is intended to reflect its true value as a dwindling resource.

Because oil meets so much of our present energy demand its price influences the costs of other forms of energy. For example some of our electricity is generated by burning oil, and fuels made from oil are used to transport coal. So the dramatic increase in the cost of oil between 1970 and 1980 has caused all energy to get more expensive, reminding us that it is far too valuable a commodity to waste.

Making the Most of Fuel

Fuel has become so precious that we cannot afford to waste a drop. Although everyone can help to reduce wastage by such measures as careful driving and home insulation a lot of useful energy is irretrievably lost before it even reaches the home, in the processes that convert fuel from one form of energy into another. The burning of coal or oil to generate electricity is an example of a particularly wasteful conversion process. Only about a third of the energy contained in the fuel actually becomes electricity – the rest is lost as heat escaping from the generators.

Some of the heat lost at power stations disappears directly into the atmosphere by escaping up the tall chimneys that provide an outlet for unwanted gases. Heat is also carried off by water pumped through a condenser to turn steam back into liquid. This water is normally discharged into a nearby river or the sea, or it may be passed through a cooling tower and then piped into the power plant again. However, the hot water can also be passed through pipes and radiators that will heat buildings, recovering energy that would otherwise just be wasted.

Water from power stations will stay hot enough to be piped underground to other buildings several miles away. About a third of the homes in Denmark are kept warm in winter by this system, which is known as district heating. However, not all district heating uses waste heat from power stations. Other projects draw their hot water supplies from boilers designed to ensure a more economic use of fuel than would be possible if every household produced its own heat. Some of these boilers are even designed to burn domestic refuse, thus tapping an extremely cheap source of energy.

As all forms of energy become more expensive, interest is growing in ways of burning fuels more efficiently. For example, scientists have now developed a more efficient way of burning coal. Crushed particles of coal and a material such as limestone are fed into a boiler and mixed together by jets of air pumped in at the base. A flame produced by oil or gas is used to set the fuel alight and the air jets agitate the mixture, making it behave like a

This infrared photograph, known as a heat scan, points to places where better home insulation is required. The colour sequence from hot to cold is yellow, red, pink, light blue, dark blue. This reveals that although the wooden end walls provide adequate insulation large amounts of heat are escaping through the brick side walls and tiled roof.

liquid. For this reason, the process is known as fluidized-bed combustion. Because the limestone does not burn, it forms a permanent 'bed' in which combustion takes place and through which water is piped to pick up the heat. The limestone also absorbs most of the coal's sulphur, which in a conventional boiler would escape into the atmosphere.

Fluidized-bed combustion is such a versatile and clean method of producing heat that it can even be used for burning finely chopped particles made from old car tyres and other forms of refuse with no risk of serious atmospheric pollution. Although the development of the system is still in its infancy, it promises to provide an attractive method of cutting fuel bills as well as disposing of some of the leftovers of modern living.

For every house heated by electricity generated at this British power station another could be heated by energy escaping from the cooling towers as steam.

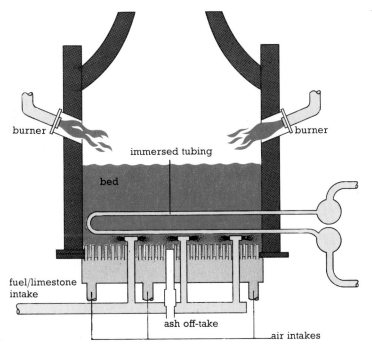

burner
immersed tubing
bed
burner
fuel/limestone intake
ash off-take
air intakes

Fluidized-bed combustion enables the efficient burning of almost any combustible substance but is particularly suited to low-grade coal with a high sulphur content. Limestone particles injected into the 'bed' along with the coal combine with up to 95 per cent of the sulphur and nitrogen pollutants during burning and can then be removed as easily handled ash.

Power for the Future?

Are we really living in an age when our energy supplies are rapidly running out? Has the world become so dependent on dwindling reserves of oil that we can never again look forward to the abundance of energy we have enjoyed in the past? Is the outlook so bleak that only those who are very rich will be able to afford adequate amounts of increasingly expensive fuel and power?

These are the kind of questions we are confronted with almost every day as newspapers and television remind us of the 'energy crisis'. What this really means is that there is a danger of a crisis arising in the future if we go on using energy as recklessly as we have in the past and we find nothing to replace oil on which we now depend for nearly half our energy needs.

Forecasting future energy demand is an extremely complicated task which involves looking for clues in what has happened in the past. For example, the total world energy demand rose nearly 50 per cent between 1925 and 1950 and increased by more than three times during the next quarter of a century up to 1975. But most of that growth was due to the relative cheapness and availability of oil. It seems unlikely that the demand for energy will ever grow in the future at anything like the rate experienced in the past. Indeed the demand for oil is now declining and we are learning to conserve energy in many ways now that all fuel is much more expensive.

The uncertainties of future energy demand make it difficult to plan nuclear power stations which may offer cheaper power than coal or oil but are very expensive to build. At present nuclear power provides a mere two per cent of world energy requirements, contributing less than half the amount produced by hydroelectric schemes. Forecasts have suggested that the amount of power produced by nuclear reactors could rise by 10–15 times by the end of the century, but those projections now seem over-optimistic. Delays have occurred in nuclear power programmes in many countries owing to widespread fear of radioactive leakages.

Many of the fears about nuclear energy may be overcome if experimental work on fusion power succeeds in reproducing the reactions which take place in the Sun to yield the abundant supplies of energy locked up in water. Whether or not fusion power can be developed on a commercial scale, the fact that the Sun is going to continue shining for billions of years provides opportunities for human ingenuity to derive energy from the movement of wind and water across the Earth's surface, or even from specially grown energy crops. However, even these renewable energy sources take time and money to develop and it is unlikely that they will contribute more than a small percentage of the world's total energy needs before the end of the century.

Whatever breakthroughs occur in the search for alternative forms of energy, it is unlikely that there will ever be a single source of energy giving us all the power we need. The world's present over-dependence on oil may have taught us a valuable lesson in this respect. The threat of an energy crisis has arisen because we have come to rely so much on oil that we have tended to forget that oil would become more expensive as supplies became scarce. The only way to avoid such costly mistakes in the future is to have a greater variety of energy sources, or perhaps better still to learn to use much less energy in our daily lives, so that readily available supplies will go a lot further.

Modern cities such as Hong Kong, shown here, represent a concentrated area of energy consumption unparalleled in history. Careful conservation of resources and a shift from oil-based power supplies will be needed if scenes such as this are to be commonplace in the twenty-first century.

Glossary

Atom: the smallest possible unit of any substance, composed of a central nucleus surrounded by electrons

Biomass: vegetable matter from which energy can be extracted

Bitumen: a heavy oil with a tar-like texture, sometimes known as pitch. It is the 'oil' ingredient of oil sands.

Blow-out: a sudden, violent gush of oil or gas which occurs when the pressure in an underground well cannot be controlled

Breeding: a nuclear process in which neutrons are released from plutonium and more plutonium is produced through the reaction of neutrons with uranium

Carbohydrates: a combination of carbon, hydrogen, and oxygen formed in plants by absorbing energy from sunlight

Celsius: a temperature scale, sometimes known as Centigrade, which is based on water having a freezing point fixed at zero and a boiling point of 100 degrees

Chain reaction: the spontaneous release of energy which occurs when sufficient nuclear fuel is assembled within a reactor core

Christmas tree: an assembly of valves connected to the surface outlet of an oil or gas well to regulate the flow

Coolant: the liquid or gas used to remove heat from the core of a nuclear reactor

Cooling pond: a water-filled trough used to store radioactive materials removed from a reactor

Core: the central part of a nuclear reactor where fuel is assembled and energy produced

Cracker: a unit in a refinery where heavy oil is broken down into lighter products

Critical mass: the minimum amount of nuclear fuel needed to trigger off a chain reaction

Crude oil: a mixture of liquid hydrocarbons

Deuterium: an atom of hydrogen with a neutron added to its nucleus. Used as a moderator and coolant in some nuclear fission reactors, and as fuel in fusion reactions.

District heating: a system that uses a central boiler or waste heat from a power station to warm nearby houses

Drilling mud: a fluid used during drilling operations that cools the bit, carries waste material to the surface, and provides weight to counteract any build-up of pressure in the well

Drillship: a vessel equipped to drill oil or gas wells in ocean exploration zones

Electrolysis: a process by which an electric current can be passed through water molecules to produce hydrogen and oxygen

Electron: the particle that forms the outer part of an atom

Energy: a measure of work done *on* a system or of work that could be done *by* a system

Fission: the splitting of atoms to release energy

Fluidized-bed combustion: a method of burning fuels, especially coal, in which the fuel is mixed with particles of limestone and agitated by jets of high-pressure air

Fossil fuel: any source of chemical energy that has been formed by burial in the Earth's crust

Fuel cell: a device in which hydrogen and oxygen combine to produce water and electricity

Fusion: the joining of the nuclear particles in light atoms to release huge quantities of energy

Gasohol: a fuel that contains alcohol derived from plants in addition to refined petroleum

Geothermal energy: heat recovered from deep within the Earth's crust

Gravity: the force by which bodies attract each other, which varies according to their mass and distance apart

Gusher: an uncontrolled flow of oil gushing from a well

Heat pump: a device, for example, a refrigerator, containing a liquid with a low boiling point which absorbs heat and gives it off elsewhere.

Hydroelectric power: electricity generated by water pressure

Hydrocarbon: the term used to describe the many combinations of hydrogen and carbon atoms which form fuels including coal, oil, and natural gas.

Laser: a device that produces a concentrated beam of light, capable of being narrowly focused

Methane: a gas formed by the combination of hydrogen and carbon. The main constituent of natural gas.

Moderator: a substance used in nuclear reactors to slow down the rate at which freed neutrons move, thus helping to control the reaction

Naphtha: a liquid refined from crude oil which can be used to produce energy gas or petrochemicals

Natural gas: a fossil fuel formed deep in the Earth's crust as a result of the effect of heat on other hydrocarbon deposits

Neutron: one of the particles in the nucleus of the atom which is released during nuclear reactions

Nuclear power: energy released when the nucleus of an atom breaks up following penetration by a neutron, or when elements such as deuterium and tritium are fused to form helium

Nucleus: the central part of an atom, which contains protons and neutrons held together by a powerful attractive force

Oil sands: deposits of liquid hydrocarbons in the form of bitumen attached to sand grains

Oil shale: oil bonded to rock deposits, which can be recovered by quarrying the rock and then roasting it

Overburden: the soil and rock that has to be removed before coal or oil sands can be extracted by quarrying

Petroleum: a fossil fuel found in underground rocks, usually known simply as oil

Photosynthesis: the process by which plants absorb energy from sunlight to combine water and carbon dioxide into carbohydrates

Pig: a device passed through oil and gas pipelines to remove wax and check that there are no obstructions

Plasma: a gas-like form of matter composed of an equal number of highly charged positive and negative particles moving independently of each other. Fusion requires extremely hot plasma.

Plutonium: a man-made element that can undergo fission to produce nuclear energy. It can be made from uranium in nuclear reactors.

Proton: one of the particles in the nucleus of the atom

Radiation: energy in the form of waves or particles

Radioactivity: radiation emitted from unstable atomic nuclei or from a nuclear reaction

Reactor: the shielded container in which nuclear reactions can be made to take place

Refinery: a processing unit where crude oil is boiled to separate out the various combinations of hydrogen and carbon atoms

Slurry: a mixture of fine particles of coal and water which can be transported by pipeline

Solar panel: a device which can be used to collect heat from sunlight and transfer it to a flow of water

Solar cell: thin layers of semi-metallic material bonded together to produce electricity when illuminated by the Sun

Strip mining: a method of recovering coal or oil sand deposits by removing the layers of soil above

Tokamak: a doughnut-shaped container surrounded by magnets and used for carrying out nuclear fusion experiments

Tritium: an atom of hydrogen that has two neutrons, produced during nuclear fusion reactions

Turbine: an engine in which propeller-like blades are driven by high-pressure gas or a moving fluid to produce mechanical energy

Uranium: a naturally occurring radioactive metal which provides the main source of fuel for nuclear reactors

Index

Credits

The Publishers gratefully acknowledge permission to reproduce the following illustrations:

Airfoto, courtesy of Energy Transportation Corporation 49*l*; Aspect Picture Library 35*l*; Barnaby's Picture Library 7*r*, 12, 23*t*, 83*l*; Billings Energy Corporation 84*l*; British Gas Corporation 44*t*, 46; British Nuclear Fuels Ltd 59; British Petroleum Photographs 31*r*, 32, 35*r*, 38; Camera Press 51; J. Allan Cash Ltd 79*r*; Central Electricity Generating Board 54, 58; Colorsport 6*l*; Photothèque E.D.F./M. Brigaud 81; Elisabeth Photo Library London Ltd 72; Esso Petroleum Company Ltd 40, 49; Robert Estall Photographs 13; E.T. Archive 87*r*; Ford Motor Co. Ltd 15*t*; Frank Frazer 37*t*, 62; Robert Harding Picture Library 78; George Hodge Associates 23*b*; Hydrogen Organisation 84*r*; Image Bank 7*l*; I.C.I. 42; Imperial Oil Ltd 39; Institute of Geological Sciences (N.E.R.C. Copyright) 29, 36; University of Reading 8; National Coal Board 21, 24, 27; Nigel Press Associates 31*l*; Picturepoint London 6*r*, 73*t*, 91; Popperfoto 11, 17; Ricardo Consulting Engineers Ltd 43; Rex Features Ltd 71; RTZ Services Library 63; Ruhr Gas 45, 48; Science Photo Library 56, 69, 90; Space Frontiers Ltd 68 (US Naval Research Laboratory), 73*b*, 83*r*; Stockphotos International 87*l*; Tony Stone Associates Ltd 47, 76; Texaco Incorporated 30; Trewin Copplestone Books Ltd 41, 55*t*; United Kingdom Atomic Energy Authority 65, 67; University of Rochester, New York 69*l*; Vision International 3 (© Heini Schneebeli), 79*l* (Paolo Koch); ZEFA 70, 93.

Front jacket photograph: Tony Stone Associates
Back jacket photograph: ZEFA.

Artwork by: Peter Berry 50; Terry Burton 16*t*, 81*t*; Kai Choi 20, 30, 32, 33, 39, 66; Keith Duran 22; Chris Forsey 76; Richard Giddon 74; Hayward and Martin Ltd 81*b*; Bill Hobson 53; Ray Martin 25*b*, 36; Wolfgang Mezger 14, 75*bl*; Michael Robinson 10, 13, 57, 59, 69, 88; Syd Roderick 25*t*; James Roper 9*t*, 26/27, 28, 29, 34, 35, 40, 60/61, 63, 64, 71, 72, 80, 82, 91; Colin Salmon 16*b*, 89; Pandora Sellars 18/19; Will Stephen 52, 54, 65, 75*r*, 84, 85, 86; Ian Stevens 9*b*; Alan Suttie 43; John Young 15.